Farm Mechanics

MACHINERY AND ITS USE TO SAVE
HAND LABOR ON THE FARM

Including

Tools, Shop Work, Driving and Driven
Machines, Farm Waterworks, Care
and Repair of Farm Implements

By

HERBERT A. SHEARER
AGRICULTURIST
Author of "Farm Buildings with Plans and Descriptions"

ILLUSTRATED WITH THREE
HUNDRED ORIGINAL DRAWINGS

CHICAGO
FREDERICK J. DRAKE & CO.
Publishers

PREFACE

More mechanical knowledge is required on the farm than in any other line
of business. If a farmer is not mechanically inclined, he is under the
necessity of employing someone who is.

Some farms are supplied with a great many handy contrivances to save
labor. Farmers differ a great deal in this respect. Some are natural
mechanics, some learn how to buy and how to operate the best farm

machinery, while others are still living in the past.

Some farmers who make the least pretensions have the best machinery and implements. They may not be good mechanics, but they have an eye to the value of labor saving tools.

The object of this book is to emphasize the importance of mechanics in modern farming; to fit scores of quick-acting machines into the daily routine of farm work and thereby lift heavy loads from the shoulders of men and women; to increase the output at less cost of hand labor and to improve the soil while producing more abundantly than ever before; to suggest the use of suitable machines to manufacture high-priced nutritious human foods from cheap farm by-products.

Illustrations are used to explain principles rather than to recommend any particular type or pattern of machine.

The old is contrasted with the new and the merits of both are expressed.

THE AUTHOR.

CONTENTS

FARM MECHANICS

CHAPTER I

THE FARM SHOP WITH TOOLS FOR WORKING WOOD AND IRON

FARM SHOP AND IMPLEMENT HOUSE

The workshop and shed to hold farm implements should look as neat and attractive as the larger buildings. Farm implements are expensive. Farm machinery is even more so. When such machinery is all properly housed and kept in repair the depreciation is estimated at ten per cent a year. When the machines are left to rust and weather in the rain and wind the loss is simply ruinous.

More machinery is required on farms than formerly and it costs more. Still it is not a question whether a farmer can afford a machine. If he has sufficient work for it he knows he cannot afford to get along without it and he must have a shed to protect it from the weather when not in use.

In the first place the implement shed should be large enough to accommodate all of the farm implements and machinery without crowding and it should be well built and tight enough to keep out the wind and small animals, including chickens and sparrows.

The perspective and plan shown herewith is twenty-four feet in width and sixty feet in length.

Figure 1.—Perspective View of the Farm Shop, Garage and Implement Shed. The doors to the right are nearly 12 feet high to let in a grain

separator over night, or during the winter, or a load of hay in case of a sudden storm.

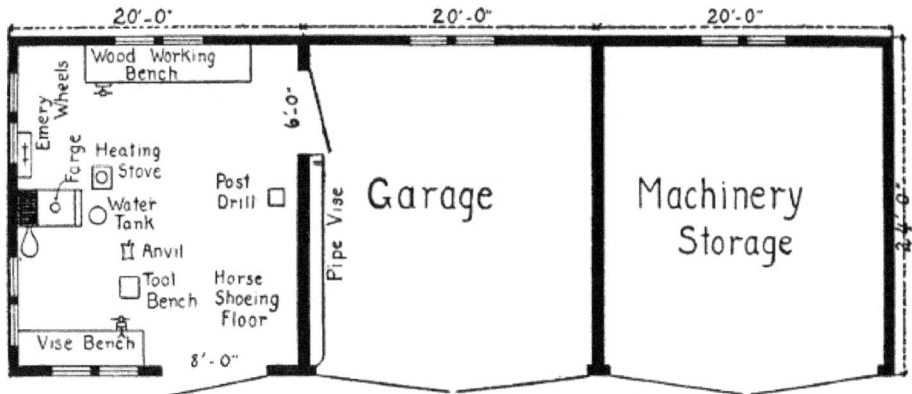

Figure 2.—Floor Plan of Shop, Garage and Storage. The building is 60 feet wide and 24 feet from front to back. The doors of the garage and tool shed are made to open full width, but 8 feet is wide enough for the shop door. All doors open out against posts and are fastened to prevent blowing shut. The work shop is well lighted and the stationary tools are carefully placed for convenience in doing repair work of all kinds. The pipe vise is at the doorway between the shop and garage so the handles of the pipe tools may swing through the doorway and the pipe may lie full length along the narrow pipe bench.

The doorways provide headroom sufficient for the highest machines, and the width when the double doors are opened and the center post removed is nearly twenty feet, which is sufficient for a binder in field condition or a two-horse spring-tooth rake.

One end of the building looking toward the house is intended for a machine shop to be partitioned off by enclosing the first bent. This gives a room twenty feet wide by twenty-four feet deep for a blacksmith shop and general repair work. The next twenty feet is the garage. The machine shop part of the building will be arranged according to the mechanical inclination of the farmer.

Figure 3.—Perspective View of Farm Implement Shed and Workshop.

A real farm repair shop is a rather elaborate mechanical proposition. There is a good brick chimney with a hood to carry off the smoke and gases from the blacksmith fire and the chimney should have a separate flue for a heating stove. Farm repair work is done mostly during the winter months when a fire in the shop is necessary for comfort and efficiency. A person cannot work to advantage with cold fingers. Paint requires moderate heat to work to advantage. Painting farm implements is a very important part of repair work.

A good shop arrangement is to have an iron workbench across the shop window in the front or entrance end of the building. In the far corner against the back wall is a good place for a woodworking bench. It is too mussy to have the blacksmith work and the carpenter work mixed up.

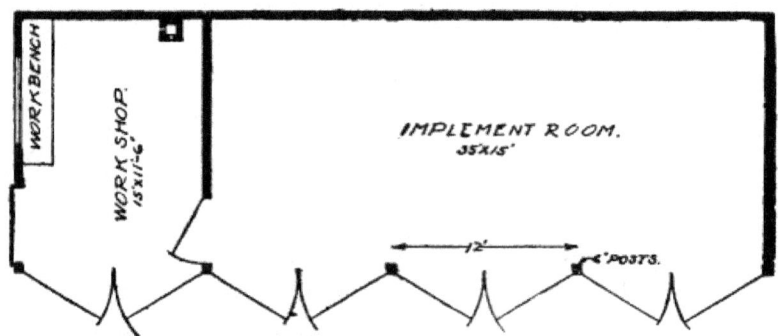

Figure 4.—Floor Plan of Farm Implement Shed, showing the workshop in one end of the building, handy to the implement storage room.

Sometimes it is necessary to bring in a pair of horses for shoeing, or to pull the shoes off. For this reason, a tie rail bolted to the studding on the side of the shop near the entrance is an extra convenience.

In a hot climate a sliding door is preferable because the wind will not slam it shut. In cold climates, hinge doors are better with a good sill and threshold to shut against to keep out the cold. Sometimes the large door contains a small door big enough to step through, but not large enough to admit much cold, when it is being opened and shut. Likewise a ceiling is needed in a cold country, while in warmer sections, a roof is sufficient. Farm shops, like other farm buildings, should conform to the climate, as well as convenience in doing the work. A solid concrete floor is a great comfort. And it is easily kept clean.

The perspective and floor plan show the arrangement of the doors, windows and chimney and the placing of the work benches, forge, anvil, toolbench and drill press.

Figures 3 and 4 show the perspective and floor plan of a farm shop and implement house 40 x 16 feet in size, which is large enough for some farms.

SHOP TOOLS

Good tools are more important on a farm than in a city workshop for the reason that a greater variety of work is required.

Measuring Mechanical Work.—In using tools on the farm the first rule should be accuracy. It is just as easy to work to one-sixteenth of an inch as to carelessly lay off a piece of work so that the pieces won't go together right.

Figure 5.—Caliper Rule. A handy slide caliper shop rule is made with a slide marked in fractions of inches as shown in the drawing. The diameter of a rivet, bolt or other round object may be taken instantly. It is not so accurate as calipers for close measurements, but it is a practical tool for farm use.

The handiest measuring tool ever invented is the old-fashioned two-foot rule that folds up to six inches in length to be carried in the pocket. Such rules to be serviceable should be brass bound. The interior marking should be notched to sixteenths. The outside marking may be laid out in eighths. The finer marking on the inside is protected by keeping the rule folded together when not in use. The coarser marking outside does not suffer so much from wear. Figure 5 shows a 12-inch rule with a slide caliper jaw.

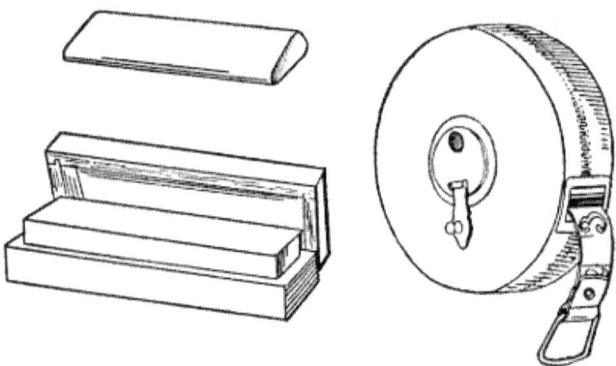

Figure 6.—Small Pocket Oilstone. Shop oilstone in a box. 100-foot measuring tapeline marked in inches, feet and rods.

In using a two-foot rule to lay off work the forward end should contain the small figures so that the workman is counting back on the rule but forward on the work, and he has the end of the rule to scribe from. In laying off a 16-foot pole the stick is first marked with a knife point, or sharp scratchawl, and try square to square one end. The work is then laid off from left to right, starting from the left hand edge of the square mark or first mark. The two-foot rule is laid flat on top of the piece of wood. At the front end of the rule the wood is marked with a sharp scratchawl or the point of a knife blade by pressing the point against the end of the rule at the time of marking. In moving the rule forward the left end is placed exactly over the left edge of the mark, so the new measurement begins at the exact point where the other left off, and so on the whole length of the stick. The final mark is then made exactly sixteen feet from the first mark.

In sawing the ends the saw kerf is cut from the waste ends of the stick. The saw cuts to the mark but does not cut it out.

In using a rule carelessly a workman may gain one-sixteenth of an inch every time he moves the rule, which would mean half of an inch in laying off a 16-foot pole, which would ruin it for carpenter work. If the pole is afterwards used for staking fence posts, he would gain one-half inch at each post, or a foot for every twenty-four posts, a distance to bother considerably in estimating acres. It is just as easy to measure exactly as it is to measure a little more or a little less, and it marks the difference between right and wrong.

WOODWORKING BENCH

In a farm workshop it is better to separate the woodworking department as far as possible from the blacksmith shop. Working wood accumulates a great deal of litter, shavings, blocks, and kindling wood, which are in the way in the blacksmith shop, and a spark from the anvil might set the shavings afire.

A woodworking bench, Figure 7, carpenter's bench, it is usually called, needs a short leg vise with wide jaws. The top of the vise should be flush with the top of the bench, so the boards may be worked when lying flat on the top of the bench. For the same reason the bench dog should lower down flush when not needed to hold the end of the board.

It is customary to make carpenter's benches separate from the shop, and large enough to stand alone, so they may be moved out doors or into other buildings.

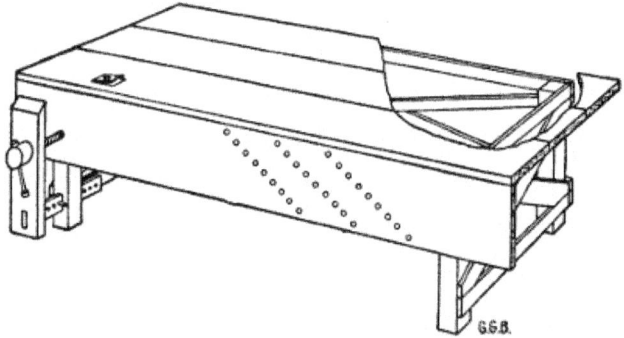

Figure 7.—Carpenter's Bench. A woodworking bench is 16' long, 3' 6" wide and 32" high. The height, to be particular, should be the length of the leg of the man who uses it. Lincoln, when joking with Stanton, gave it as his opinion that "a man's legs should be just long enough to reach the ground." But that rule is not sufficiently definite to satisfy carpenters, so they adopted the inside leg measurement. They claim that the average carpenter is 5' 10" tall and he wears a 32" leg.

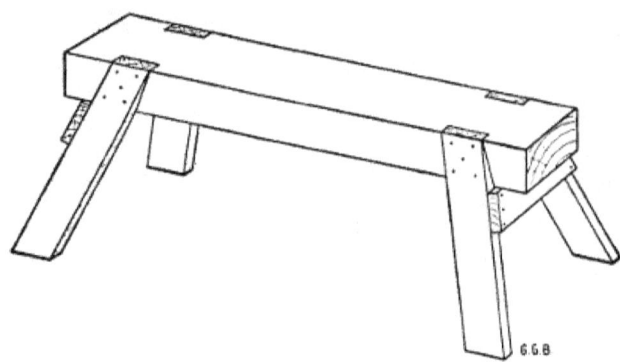

Figure 8.—Carpenter's Trestle, or Saw-Bench. The top piece is 4 x 6 and the legs are 2 x 4. There is sufficient spread of leg to prevent it from toppling over, but the legs are not greatly in the way. It is heavy enough to stand still while you slide a board along. It is 2 feet high.

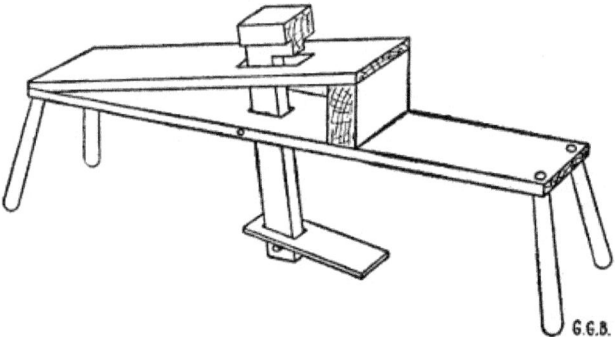

Figure 9.—Shave Horse. For shaping pieces of hardwood for repair work. A good shave horse is about 8′ long and the seat end is the height of a chair. The head is carved on a hardwood stick with three projections to grip different sized pieces to be worked.

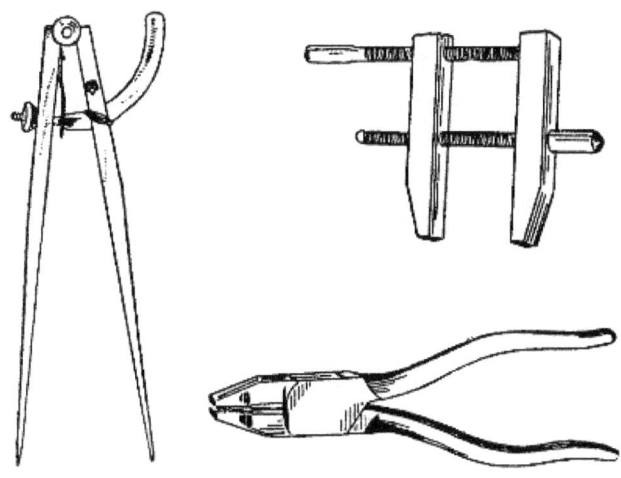

Figure 10.—Compasses, Wooden Clamp and Cutting Pliers.

Carpenter benches may be well made, or they may be constructed in a hurry. So long as the top is true it makes but little difference how the legs are attached, so long as they are strong and enough of them. A carpenter bench that is used for all kinds of work must be solid enough to permit hammering, driving nails, etc. Usually the top of the bench is straight, true and level and it should be kept free from litter and extra tools.

Good carpenters prefer a tool rack separate from the bench. It may stand on the floor or be attached to the wall. Carpenter tools on a farm are not numerous, but they should have a regular place, and laborers on the farms should be encouraged to keep the tools where they belong.

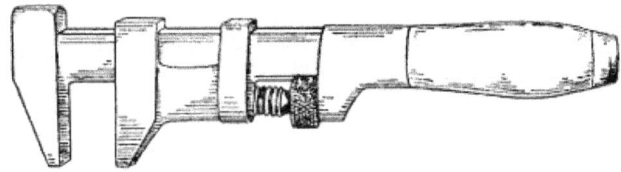

Figure 11.—Monkey-Wrenches are the handiest of all farm wrenches, but they were never intended to hammer with. Two sizes are needed—an eight-inch for small nuts and a much larger wrench, to open two inches or more, to use when taking the disks off the shafts of a disk harrow. A large pipe-wrench to hold the round shaft makes a good companion tool for this work.

WOODWORKING TOOLS

Every farmer has an axe or two, some sort of a handsaw and a nail hammer. It is astonishing what jobs of repair work a handy farmer will do with such a dearth of tools. But it is not necessary to worry along without a good repair kit. Tools are cheap enough.

Such woodworking tools as coarse and fine toothed hand saws, a good square, a splendid assortment of hammers and the different kinds of wrenches, screw clamps, boring tools—in fact a complete assortment of handy woodworking tools is an absolute necessity on a well-managed farm.

The farm kit should contain two sizes of nail hammers, see Figure 15, one suitable to drive small nails, say up to eight penny, and the other for large nails and spikes; a long thin-bladed handsaw, having nine teeth to the inch, for sawing boards and planks; a shorter handsaw, having ten teeth to the inch, for small work and for pruning trees. A pruning saw should cut a fine, smooth kerf, so the wound will not collect and hold moisture.

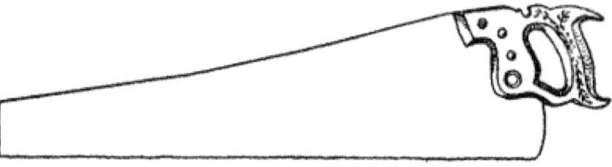

Figure 12.—Hand Saw. This pattern, both for cross cut and rip saw, has been adopted by all makers of fine saws. Nine teeth to the inch is fine enough for most jobs on the farm.

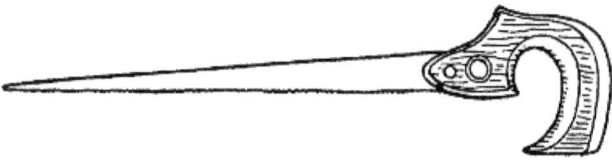

Figure 13.—Keyhole Saw with point slim enough to start the cut from a half-inch auger hole.

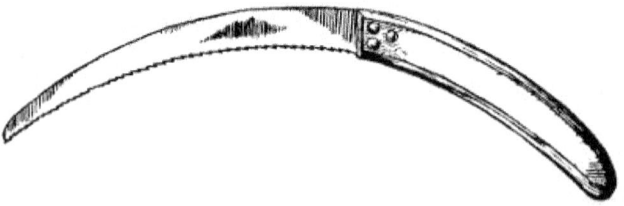

Figure 14.—Bramble Hook for trimming berry bushes and cleaning out

fence corners. It has a knife-edge with hooked sawteeth.

Farmers' handsaws are required to do a great many different kinds of work. For this reason, it is difficult to keep them in good working condition, but if both saws are jointed, set and filed by a good mechanic once or twice a year, they may be kept in usable condition the rest of the time by a handy farm workman, unless extra building or special work is required.

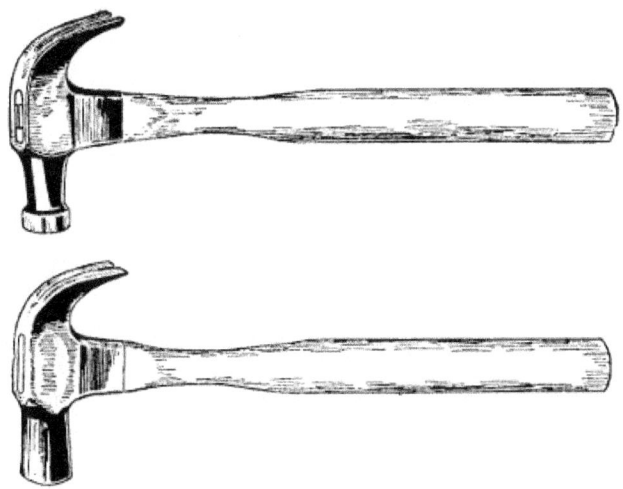

Figure 15.—Nail Hammers. Two styles. The upper hammer is made with a ball peen and a round face. It is tempered to drive small nails without slipping and shaped to avoid dinging the wood. This hammer should weigh 18 or 19 ounces, including the handle. The lower hammer is heavier, has a flat face and is intended for heavy work such as driving spikes and fence staples.

A long-bladed ripsaw is also very useful, and what is commonly termed a keyhole saw finds more use on the farm than in a carpenter's shop in town. It is necessary frequently to cut holes through partitions, floors, etc., and at such times a keyhole saw works in just right.

Handaxes are necessary for roughing certain pieces of wood for repair jobs. Two sizes of handaxes for different kinds of work are very useful, also a wide blade draw shave, Figure 16, and shave horse, Figure 9. A steel square having one 24-inch blade and one 18-inch is the best size. Such squares usually are heavy enough to remain square after falling off the bench forty or fifty times. A good deal depends upon the quality of the steel.

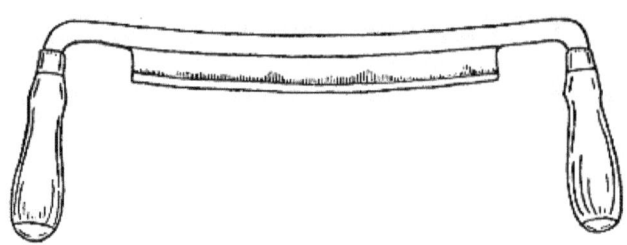

Figure 16.—Drawing-Knife with wide blade for finishing straight surfaces.

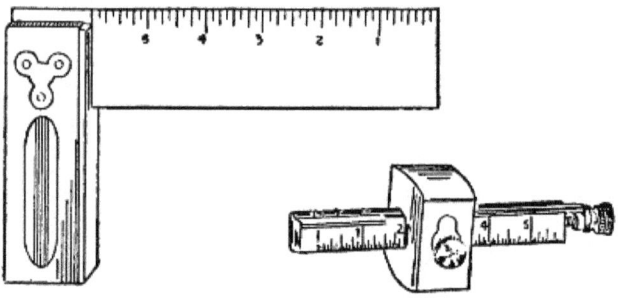

Figure 17.—Try-Square With Six-Inch Blade. Wood, brass and steel are the proper materials for a try-square. A double marking gauge for scribing mortises is also shown.

Steel squares differ in the measuring marks, but the kind to buy has one side spaced to sixteenths and the other side to tenths or twelfths. The sixteenth interest farmers generally, so that special attention should be given this side of the square. The lumber rule on some squares is useful, but the brace rules and mitre calculations are not likely to interest farmers.

Screw-drivers should be mostly strong and heavy for farm work. Three sizes of handled screw-drivers of different lengths and sizes, also two or three brace bit screw-drivers are needed. One or two bits may be broken or twisted so the assortment is sometimes exhausted before the screw is started.

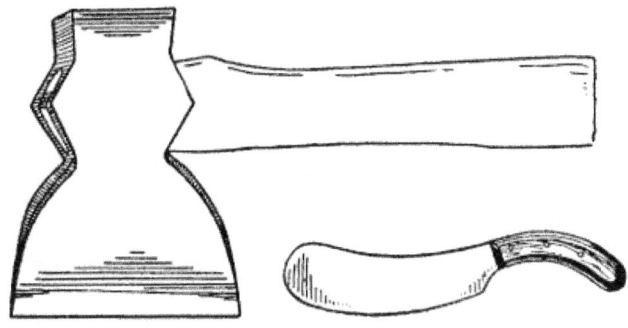

Figure 18.—Heavy Hand Axe for Use on the Shop Chopping Block. A beet topping knife is shown also.

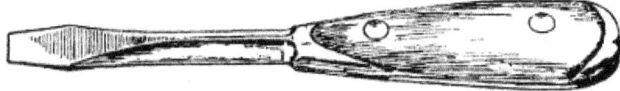

Figure 19.—Heavy Screwdriver. The strongest and cheapest screwdriver is made from a single bar of steel. The wooden handle is made in two parts and riveted as shown.

Pinch bars and claw bars are very useful in a farm tool kit. Farm mechanical work consists principally in repairing implements, machinery, fences and buildings. Always a worn or broken part must be removed before the repair can be made. A pinch bar twenty-four inches long, Figure 21, with a cold chisel end, and another bar eighteen inches long with a crooked claw end, Figure 22, for pulling nails and spikes comes in very handy. These two bars should be made of the best octagon steel, seven-eighths of an inch in diameter.

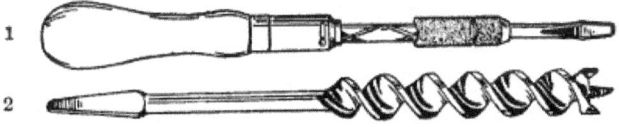

Figure 20.—(1) Ratchet Screwdriver. It does rapid work and will last a generation if carefully used. (2) Auger-Bit of the Side Cutter Type. A full set is needed. They are not for boring into old wood. Running once against a nail ruins one of these bits.

Figure 21.—Handspike. A wooden handspike or pry is about seven feet

long by 3 inches thick at the prying end. In the North it is usually made from a hickory or an ironwood or a dogwood sapling. The bark is removed and the handle is worked round and smooth on the shave horse. It is better to cut the poles in the winter when the sap is in the roots. After the handspikes are finished they should be covered deep with straw so they will season slowly to prevent checking.

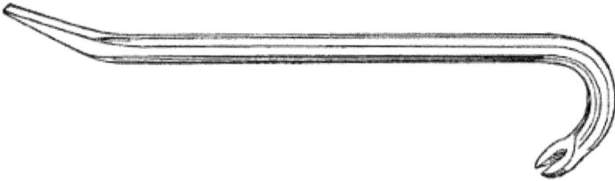

Figure 22.—Wrecking Bar for pulling nails and to pry broken parts from other wreckage.

Figure 23.—Carpenter's Level. For practical farm work the level should be 24" or 30" long. Wood is the most satisfactory material. The best levels are made up of different layers of wood glued together to prevent warping or twisting. For this reason a good level should be carefully laid away in a dry place immediately after using.

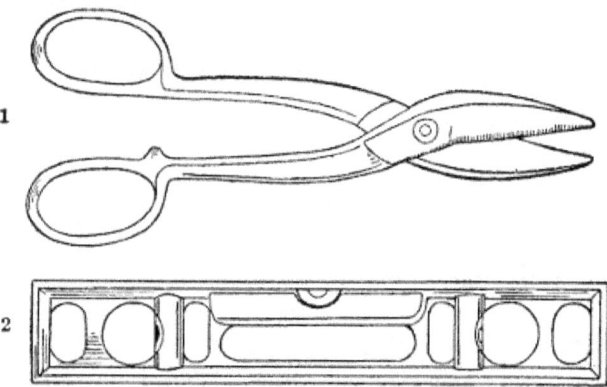

Figure 24.—(1) Snips for cutting sheet metal. (2) Carpenter's Level, iron stock.

Figure 25.—Wood-Boring Twist Drill Bit. Twist drills for wood have longer points than drills for boring iron.

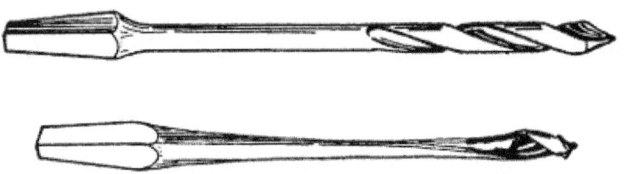

Figure 26.—Pod-Bit. The fastest boring gimlet bits are of this pattern. They are made in sizes from to $\frac{1}{8}''$ to $\frac{3}{8}''$ and are intended for boring softwood.

Figure 27.—Auger-Bits. For smooth boring the lip bits are best. The side cutters project beyond the cutting lips to cut the circle ahead of the chips. For boring green wood the single-worm clears better than the double-worm bit.

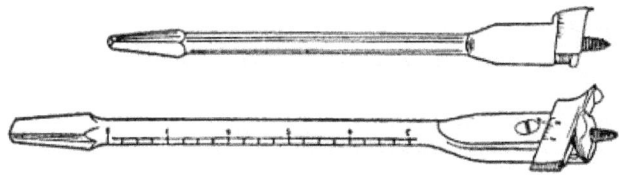

Figure 28.—Extension Boring Bits. The cutting lips may be set to bore holes from $\frac{1}{2}''$ to $3''$ in diameter. They are used mostly in softwood.

Figure 29.—Ship Auger. This shape auger is made with or without a screw point. It will bore straighter in cross-grained wood without a point.

Figure 30.—Long Ship Auger.

Figure 31.—Bridge Auger. The long handle permits the workman to stand erect while boring. The home made handle is welded onto the shank of a ship auger.

A wooden carpenter's level, Figure 23, two feet long, with a plumb glass near one end, is the most satisfactory farm level, an instrument that is needed a great many times during the year.

Good brace bits are scarce on farms. They are not expensive, but farmers are careless about bits and braces. Two sizes of braces are needed, a small brace for small pod bits and twist drills, and a large ratchet brace with a 6-inch crank radius for turning larger bits.

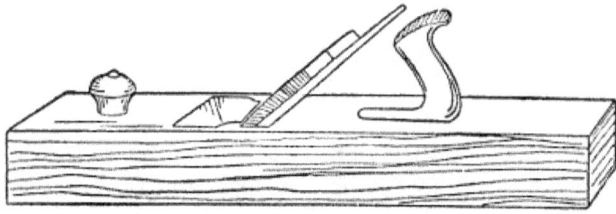

Figure 32.—Carpenter's Jointer.

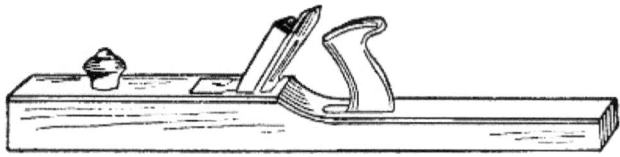

Figure 33.—Fore-Plane. This style plane is preferred to a regular jointer for most farm work.

Twist drill bits will bore both wood and iron, and they are not expensive up to three-eighths inch or one-half inch. But for larger sizes from one-half inch to one inch the finest lip wood boring bits will give the best satisfaction. Extension bits are used for boring holes larger than one inch. Two extension bits are better than one bit with two lip cutters. They will bore holes in soft wood in sizes from one inch to three inches.

Other cutting tools such as jack plane jointer and smoothing plane, also an assortment of chisels, belong to the farm equipment.

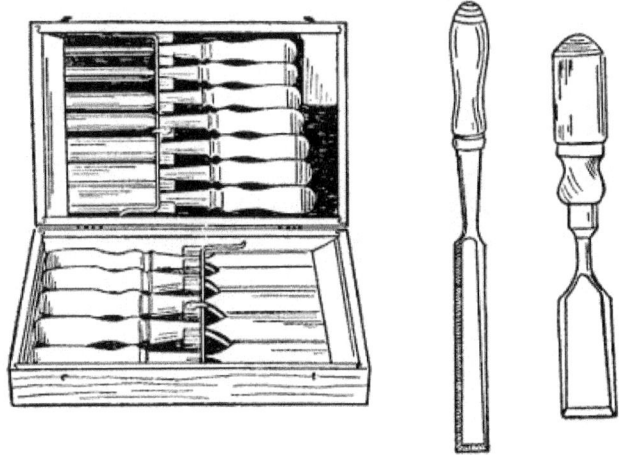

Figure 34.—Tool Box of Socket Chisels and Gouges. The chisels are sized from ½" to 2" in width. The two chisels to the right show different patterns.

All cutting tools should be of the best design and the best steel. If they are properly used and taken care of, the different jobs of repair work can be handled quickly and to great advantage.

FARM GRINDSTONE

A grindstone may be gritty without being coarse so it will bite the steel easily and cut it away quickly. A good stone is a very satisfactory farm implement, but a greasy stone is a perpetual nuisance.

There are grindstones with frames too light. The competition to manufacture and sell a grindstone for farm use at the cheapest possible price has resulted in turning out thousands of grindstone frames that possess very little stability.

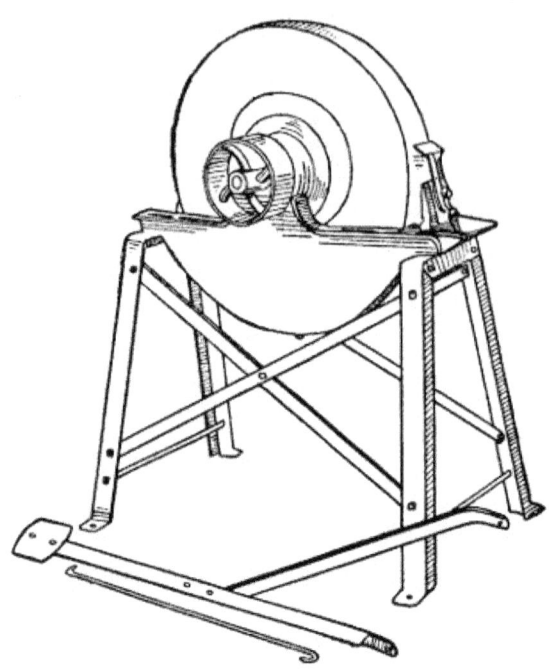

Figure 35.—Grindstone. The speed of a grindstone varies with the diameter of the stone. It should turn just fast enough to keep a flow of water on the upper face surface. If the stone turns too slow the water will run down; if too fast, it will fly off.

Grindstones should be kept under cover; the best stone will be injured by leaving it in the hot sun. The sun draws the moisture out of the upper side and leaves the lower side damp and soft so that in use the stone soon becomes flat sided. The wet side freezes in winter, which is a disintegrating process.

The best stones, with good care, will become uneven in time. The remedy is to true them with a quarter-inch soft iron round rod used like a lathe tool over an iron rest placed close to the stone on a level with the center of the stone. The rod is held against the stone in such a way as to cut away the high bumps and make the stone truly round. The stone cuts away best when it is dry. A small rod is better than a large rod. It digs into the stone better and takes out a deeper bite. Large power stones in machine shops are trued up in this way frequently. Farm stones often are neglected until they wabble so badly that it is difficult to grind any tool to an edge. If the grindstone is turned by a belt from an engine the work of truing may be done in a few minutes. If the stone is turned by hand the work of making it round takes longer and requires some muscle, but it pays.

The face of a grindstone should be rounded slightly, and it should be kept so by grinding the tools first on one side of edge of the stone, then on the other, with the cutting edge of the tool crosswise to the face of the stone.

For safety and to prevent a sloppy waste of water the stone should turn away from the operator.

The best way to keep a stone moist is by a trickle of water from an overhead supply. Troughs of water suspended under the stone are unsatisfactory, because the water soon gets thick and unfit for use. Such troughs are forgotten when the job is done, so that one side of the stone hangs in the water. An overhead supply of water leaks away and no damage is done.

Grindstone frames are best made of wood 3″ x 4″ thoroughly mortised together and well braced with wooden braces and tied across with plenty of iron rods. A good grindstone frame could be made of angle iron, but manufacturers generally fail in the attempt.

There are good ball-bearing grindstone hangers on the market, both for hand crank stones and for belt use.

The belt is less in the way if it is brought up from below. This is not difficult to do. A grindstone turns slower than any other farm machine so a speed reducing jack may be bolted to the floor at the back of the grindstone a little to one side to escape the drip. This arrangement requires a short belt but it may have the full face width of the pulley as the tight and loose pulleys are on the jack shaft.

Emery Grinders.—There are small emery wheels made for grinding disks that work quickly and cut an even bevel all around. They are made in pairs and are attached to the ends of a mandrel supported by a metal stand which is bolted to a bench. The same rig is used for sickle grinding and other farm jobs.

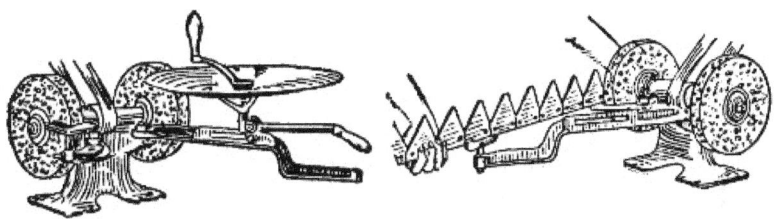

Figure 36.—Emery Grinder. The illustrations show two kinds of grinding

that double emery wheels are especially adapted to. To grind a mowing-machine knife it is necessary to reverse. By placing the rest opposite the center between the two wheels the bevel will be the same on both sides, or edges, of the section.

BLACKSMITH SHOP

The furniture in a blacksmith shop consists of forge, anvil, half barrel, vise bench, drill press and tool rack. A farm shop also has a heating stove, shave horse, a woodworking bench, a good power driven grindstone and a double emery grinder.

Forge.—The old-fashioned forge laid up with brick in connection with an old-fashioned chimney is just as popular as ever. The same old tuyer iron receives the air blast from the same old style leather bellows, and there is nothing more satisfactory. But there are modern portable forges, Figure 37, made of iron, that are less artistic, cheaper, take up less room and answer the purpose just about as well. The portable iron forge has a small blower attached to the frame which feeds oxygen into the fire. There are a good many different sizes of portable forges. Most of them work well up to their advertised capacity.

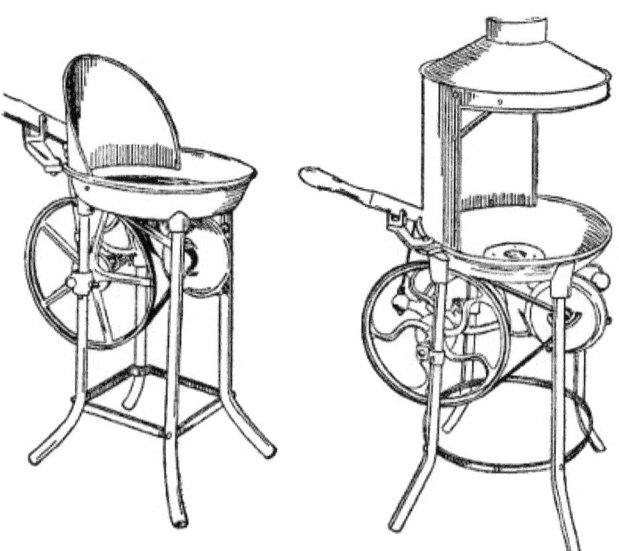

Figure 37.—Portable Forges. The smaller forge is for light work such as heating rivets for iron bridge construction. The larger forge to the right is meant for blacksmith work.

Generally, farm forges are not required to develop a great amount of heat. Farmers do but little welding, most of the forge work on the farm being confined to repair work such as heating brace irons, so they may be easily bent into the proper shape, or to soften metal so that holes may be punched through it easily.

Sharpening harrow teeth, drawing out plow points and horseshoeing are about the heaviest forge jobs required in a farm blacksmith shop, so that a medium size forge will answer the purpose.

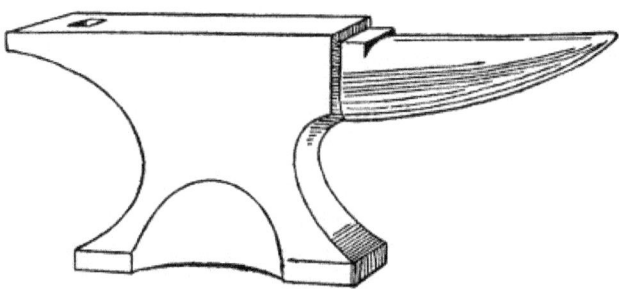

Figure 38.—Anvil. The only satisfactory anvil is forged out of ingot steel with a power trip-hammer. It should weigh 140 pounds.

Anvil.—An anvil should weigh at least 120 pounds; 140 is better. It should be set six feet from the center of the fire to the center of the anvil. It should be placed on a timber the size of the base of the anvil set three feet in the ground. The top of the anvil should be about thirty inches high. Holmstrom's rule is: "Close the fist, stand erect with the arm hanging down. The knuckles should just clear the face of the anvil."

Bench and Vise.—The vise bench should be made solid and it should face a good light. The bench window should look to the east or north if possible. It should be about four feet high and eight feet long, with the window sill about six inches above the bench.

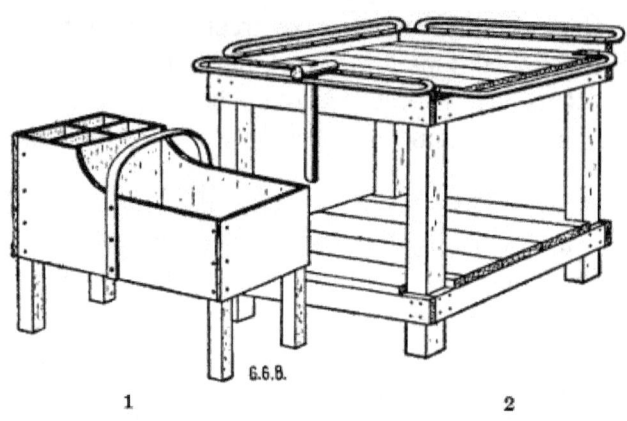

1   2

Figure 39.—(1) Shoeing Tool Box. The four small compartments are for horseshoe nails of different sizes. There may be a leather loop for the paring knife. The low box end is for the shoeing hammer, rasp, nippers and hoof knife. (2) Blacksmith Tool Rack. Tongs, handled punches and cutters are hung on the iron rails. Hammers are thrown on top. The lower platform is the shop catch-all.

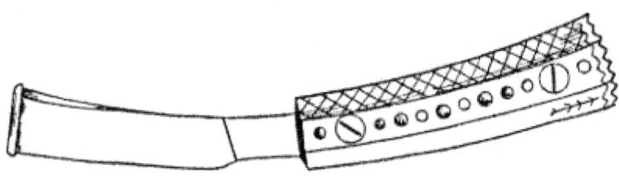

Figure 40.—Shoeing Knife. Good temper is the main qualification. All shoeing knives are practically the same shape, although they may vary in size.

Two and one-half feet is the usual height for a workbench above the floor. The best workbench tops are made by bolting together 2 x 4s with the edges up. Hardwood makes the best bench, but good pine will last for years. The top surface should be planed true and smooth after the nuts are drawn tight.

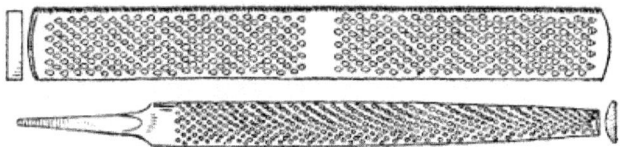

Figure 41.—Horseshoeing Rasp and Wood Rasp. These are necessary tools in the farm shop.

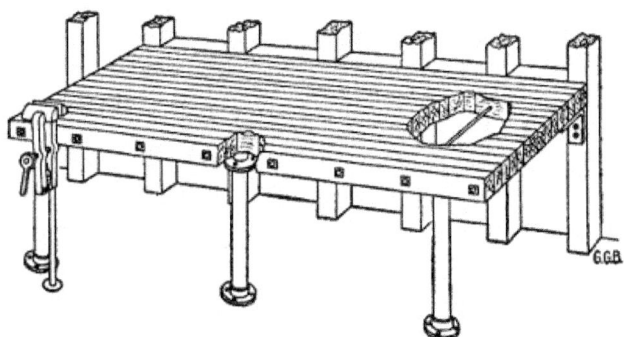

Figure 42.—Iron Work Bench. Solid is the first specification for an iron shop bench. It should be three feet wide, not less than eight feet long and about 32 inches high. The top is made of 2 x 4s placed on edge and bolted together. The supports are 2 x 6 bolted to the shop studding and braced back to the studding at the sill. The front part of the bench is supported by iron legs made of gas-pipe with threaded flanges at top and bottom. Heavy right angle wrought iron lugs are used to fasten the top of the bench to the studding. The foot of the vise leg is let into the floor of the shop or into a solid wooden block sunk in the ground.

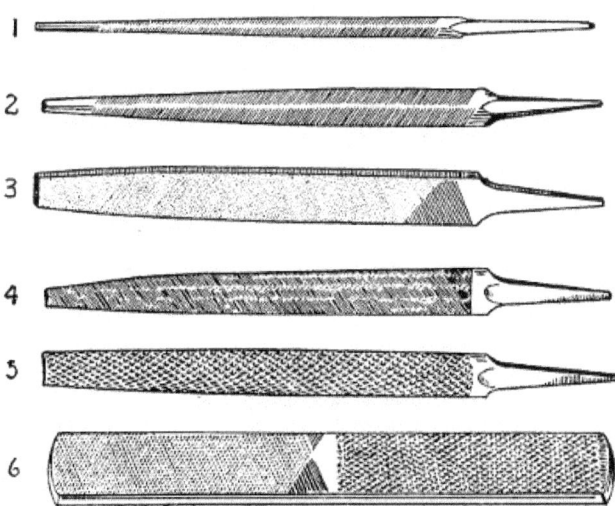

Figure 43.—Assortment of Files and Rasps needed in a farm shop. (1) Slim three-cornered handsaw-file. (2) Common three-cornered file suitable for filing a buck-saw. (3) Double-cut, or bastard, 10-inch flat file. (4) Single-cut, or mill file, either 10 or 12 inches.

(5) Half-round 10-inch wood rasp. (6) Horseshoer's rasp.

Figure 44.—File Handle. Basswood makes the most satisfactory file handles. They are fitted by carefully turning them onto the file shank to take the right taper. There should be a handle for each file. The handle should be the right size and fitted straight with the file so the file will take the same angle to the work when turned over.

Figure 45.—Nail Set. On all wooden surfaces to be painted nails should be carefully driven with a round peen nail hammer and the heads sunk about one-eighth of an inch deep with a nail set. The holes may then be filled with putty and covered smoothly with paint.

Figure 46.—Cold-Chisel. There are more flat cold-chisels than all other shapes. They are easily made in the farm shop and it is good practice. They are usually made from octagon steel. Different sizes are needed according to the work in hand. A piece of ⅝" steel 6" long makes a handy cold-chisel for repair work.

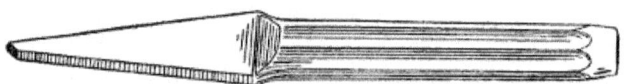

Figure 47.—Cape Cold-Chisel. It may be tapered both ways or one way to a cutting edge, or one edge may be rounded.

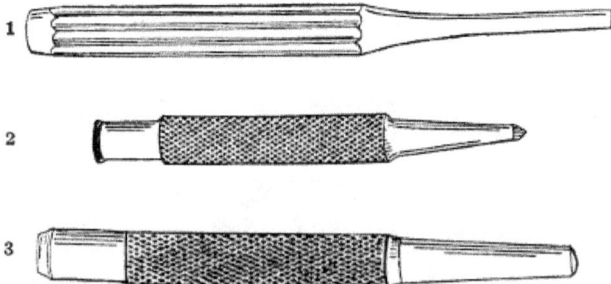

Figure 48.—(1) Tinner's Punch. Made of octagon steel in sizes to fit the rivets. The cutting end is flat and has sharp edges made by roll filing. It should be about 7" long and

from ⅜″ to ½″ in diameter, according to the size of rivet and thickness of sheet metal to be punched. (2) Prick Punch. Usually made rather short and stocky. It may be ½″ or ⅝″ diameter and 4½″ to 5″ long. (3) Hot-iron Punch. Made in many sizes and lengths. The taper should be the same as the drawing.

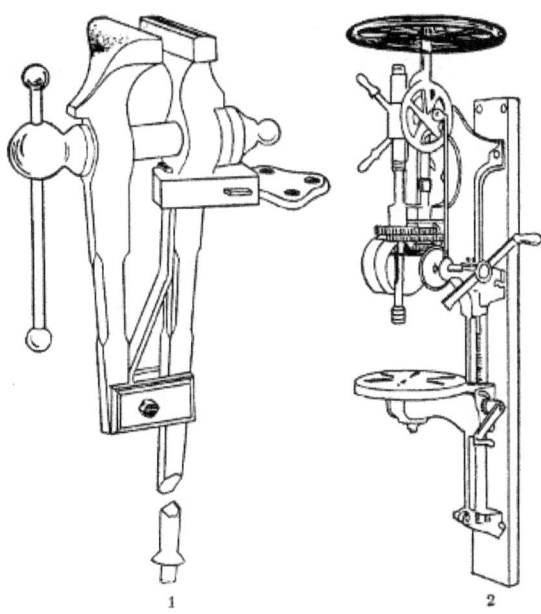

1   2

Figure 49.—(1) Blacksmith Vise. The old-fashioned leg vise is the most satisfactory for the blacksmith shop. It should have 5″ jaws. (2) Power Post Drill. Belt power is practical for the post drill in a farm shop. The hand crank may be easily attached when needed.

The bench vise should be heavy. A vise is used for bending iron hot from the forge. Unless the jaws are large, the hot iron is likely to heat the vise sufficiently to draw the temper. Heavy jaws are solid enough to [36-37-38] support the iron when it is being hammered. Often heavy hammers are used for this purpose. A heavy vise holds the work solid, because it may be screwed so much tighter than a light vise. A heavy vise will hold light work, but a light vise will not hold heavy work. Heavy vises cost more, but they are cheaper in the end and more satisfactory at all times. A leg vise with five-inch jaws weighs about sixty pounds; five and one-half-inch jaws, eighty pounds. A machinist's vise is made to bolt on top of the bench. It will answer for blacksmith work on the farm, but is not as good as the old-fashioned leg vise. A machinist's vise is very useful in the garage, but it would hardly be necessary to have two heavy vises. The pipe vise belongs on a separate bench, which may be a plank bracketed against the side of the room.

Drill-Press.—The most satisfactory drill-press for use on a farm is the upright drill that bolts to a post. There is usually a self feed which may be regulated according to the work. The heavy flywheel keeps the motion steady, and because there is no bench in the way, wagon tires may be suspended from the drill block, so they will hang free and true for drilling. Often long pieces of straight iron are drilled with holes spaced certain distances apart. It is easier to pass them along when they lie flat side down on the drill block. To use a drill properly and safely, the chuck must run true. It is easy to break a drill when it wabbles.

Most drills are made on the twist pattern, and it is something of a trick to grind a twist drill, but anyone can do it if he tackles the job with a determination to do it right. In grinding a twist drill, use a new drill for pattern. Grind the angles the same as the new drill, and be careful to have the point in the center. A little practice will make perfect.

Mechanics will say that no one except an expert should attempt to grind a twist drill, but farmers who are mechanically inclined are the best experts within reach. It is up to a farmer to grind his own drills or use them dull.

In drilling wrought iron either water or oil is required to cool the drill, but cast iron and brass are drilled dry. Light work such as hoop-iron may be drilled dry, but the cutting edge of the drill will last longer even in light work if the drill is fed with oil or water.

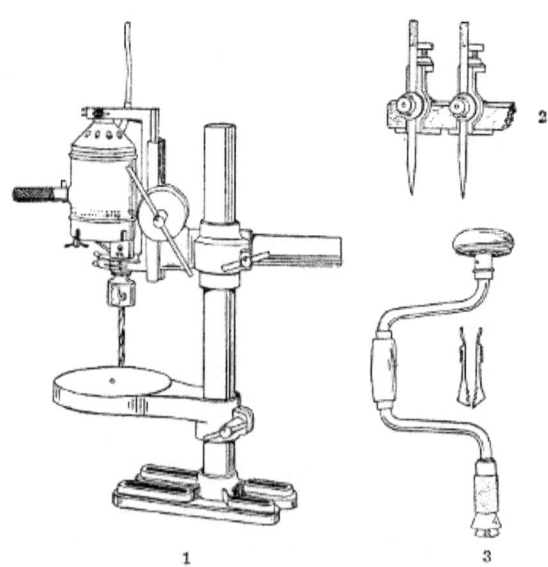

Figure 50.—(1) Electric Drill-Press. A small electric motor is attached to the drill spindle. (2) Tram Points. Two steel points are fitted with thumbscrew clamps to fasten them to a

long wooden bar. They are used to scribe circles too large for the compasses. (3) Ratchet-Brace. Two braces, or bitstocks, are needed. A large brace with a 6″ radius for large bits and a small brace with a 3″ or 3½″ radius for small bits.

In using drill-presses, some extra attachments come in very handy, such as a screw clamp to hold short pieces of metal. Before starting the drill, a center punch is used to mark the center of the hole to be bored and to start the drill in the right spot.

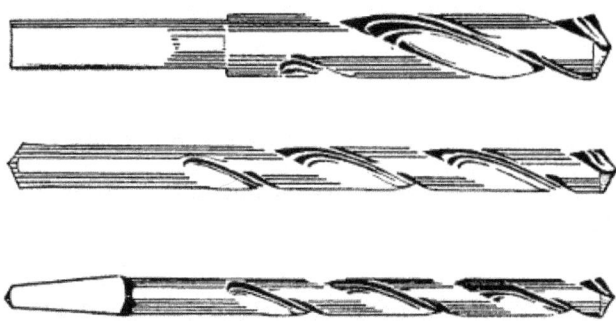

Figure 51.—Twist-Drills. Round shank for the post drill and square taper shank for brace work. Brace drills are small, ¼″ or less.

Figure 52.—Taper Reamer. Used to enlarge, or true, or taper a hole that has been drilled or punched.

Figure 53.—Another style of Reamer.

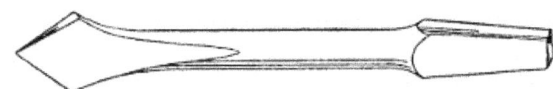

Figure 54.—Countersink. This is the old style, blacksmith-made, flat countersink. It will do quick work but not so smooth as the fluted kind.

In doing particular work, the drill may be re-centered when it starts wrong. This is done with a small round-nosed cold chisel. If the work is not very particular, the drill may be turned a little to one side by slanting

29

the piece to be drilled. This plan is only a makeshift, however, the proper way being to block the work level, so that the drill will meet it perpendicularly. However, by starting carefully, the hole may be bored exactly as required.

Iron Working Tools.—Forge tools for a farm shop need not be numerous. Several pairs of tongs, one blacksmith hammer, one sledge, one hardy, one wooden-handled cold chisel, one pair pincers, one paring knife, one shoeing rasp, and one shoeing hammer will do to begin with.

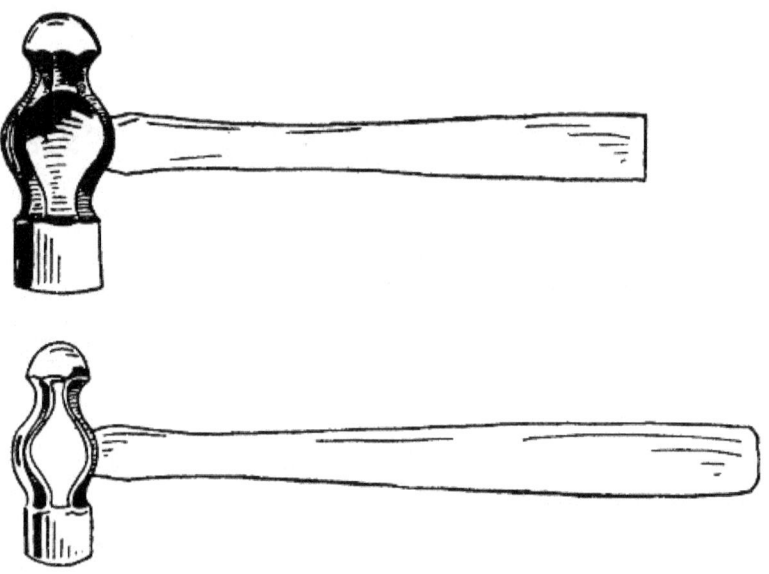

Figure 55.—Machinist's Hammers. A medium weight should be selected for farm repair work. It should be hung so the end of the handle clears half an inch when the face rests flat on the bench.

Monkey-wrenches come first in the wrench department. The farmer needs three sizes, one may be quite small, say six inches in length, one ten inches, and the other large enough to span a two-inch nut. And there should be an ironclad rule, never use a monkey-wrench for a hammer. For work around plows, cultivators, harvesters, and other farm machines, a case of S wrenches will be greatly appreciated. Manufacturers include wrenches with almost all farm machines, but such wrenches are too cheap to be of much use.

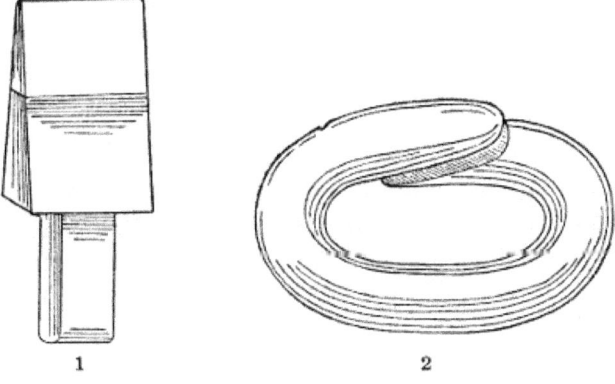

Figure 56.—(1) Hardy. The anvil hardy is used more than any other anvil tool except the blacksmith's hammer and tongs. (2) A Cold-Shut Link that may be welded, riveted or simply pounded shut.

Figure 57.—Calipers: (1) A pair of tight-joint inside calipers. (2) Its mate for taking outside dimensions. (3) A pair of spring-jointed, screw-adjustment inside calipers for machinists' use.

Figure 58.—Blacksmith Tongs. Straight tongs made to hold ⅜" iron is the handiest size. Two or three pairs for larger sizes of iron and one pair smaller come in handy.

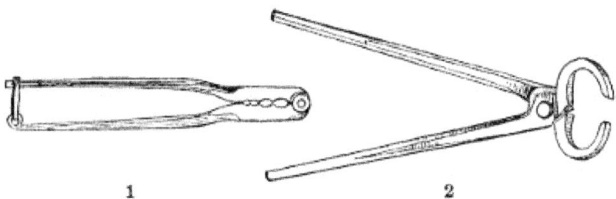

Figure 59.—(1) Wire Splicer. The oval openings in the tool are of different sizes. They are made to hold two wires, close together, with ends projecting in opposite directions. Each end is wound around the other wire. The ends are then notched with a three-cornered file

and broken off short and filed smooth. The splicing tool should be thin, about $\frac{1}{8}''$ or $\frac{3}{16}''$, to bring the two twists close together. This is especially necessary in making hoops for wooden pails. (2) Blacksmith Shoeing Pincers, used to pull horseshoes. They should close together to catch a nail by the head.

For heavier work pipe-wrenches are absolutely necessary. The reason for having so many wrenches is to save time when in the field. It often happens that men and horses stand idle waiting for what should be a quick repair job.

Figure 60.—(1) Cotter Pin Tool. Handy for inserting or removing all sorts of cotter keys. (2) Nest of S Wrenches of different sizes. Farmers have never appreciated the value of light, handy wrenches to fit all sorts of nuts and bolt heads closely.

For bench work a riveting hammer and a ball peen machinist's hammer are needed. A nest of S wrenches, two rivet sets, cold chisels, round punches and several files also are required.

The same twist drills up to three-eighths-inch will do for iron as well as wood. However, if much drilling is done, then round shank twist drills to fit the drill chuck will work better. Farmers seldom drill holes in iron larger than one-half inch. For particular work, to get the exact size, reamers are used to finish the holes after drilling. Screw holes in iron are countersunk in the drill-press.

Figure 61.—Hack Saw. One handle and a dozen blades. The frame should be stiff enough either to push or pull the saw without binding. The teeth may point either way to suit the work in hand.

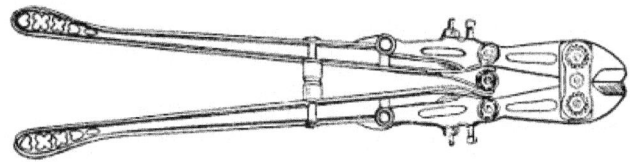

Figure 62.—Powerful Bolt Cutter. It is intended for factory use.

For small work, twist drills with square shanks for brace use should range in sizes from one thirty-second of an inch up to one-quarter inch, then every one-sixteenth inch up to one-half inch.

For boring screw holes in wood the quickest work is done with pod bits. Not many sizes are needed, but they are cheap, so that a half dozen, ranging from one-sixteenth to one-quarter inch or thereabouts, will be found very useful. Pod bits belong to the wood department, but on account of being used principally for screw sinking, they are just as useful in the iron working department as in the carpenter shop.

Sheet metal snips for cutting sheet metal properly belong with the iron working tools. Snips are from ten to fourteen inches in length. A medium size is best for miscellaneous work. If kept in good working order twelve-inch snips will cut 18-gauge galvanized or black iron. But a man would not care to do a great deal of such heavy cutting.

Figure 63.—Cutting Nippers. For cutting the points from horseshoe nails after they are driven through the hoof to hold the shoe in place. These nippers are hard tempered and should not be used for any other purpose.

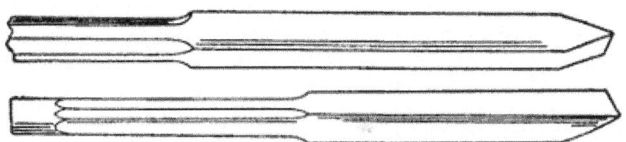

Figure 64.—Two Shapes of Steel Crowbars.

Pipe-Fitting Tools.—Recent farm improvements require a few tools that rightfully belong to plumbers. Every farm has some kind of water supply for domestic use and for live-stock. A great many farm machines require pipe tools for repair work. Every year more plumbing reaches the farm.

Plumbing work is no more difficult than other mechanical work, if the tools are at hand to meet the different requirements. One job of plumbing that used to stand out as an impossibility was the soldering together of lead pipes, technically termed "wiping a joint." This operation has been discontinued. Every possible connection required in farm plumbing is now provided for in standardized fittings. Every pipe-fitting or connection that conducts supply water or waste water nowadays screws together. Sizes are all made to certain standards and the couplings are almost perfect, so that work formerly shrouded in mystery or hidden under trade secrets is now open to every schoolboy who has learned to read.

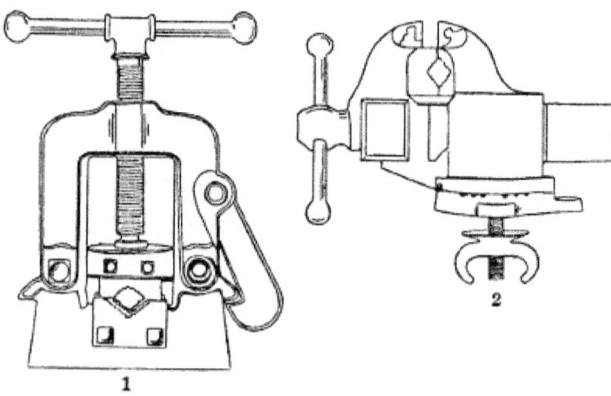

Figure 65.—(1) Pipe Vise. Hinged to open for long pipes. (2) Machinist's Vise. Made with a turntable to take any horizontal angle. The pipe jaws are removable.

The necessary outfit to handle all the piping and plumbing on the farm is not very expensive, probably $25.00 will include every tool and all other appliances necessary to put in all the piping needed to carry water to the watering troughs and to supply hot and cold water to the kitchen and the bathroom, together with the waste pipes, ventilators and the sewer to the septic tank. The same outfit of tools will answer for repair work for a lifetime.

Farm water pipes usually are small. There may be a two-inch suction pipe to the force pump, and the discharge may be one and a half inch. But these pipes are not likely to make trouble.

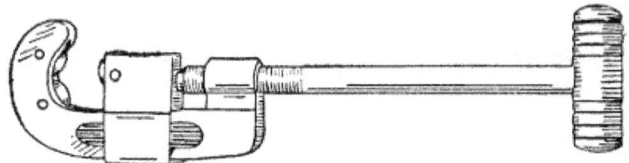

Figure 66.—Pipe Cutter. The most satisfactory pipe cutter has three knife-edge roller cutters which follow each other around the pipe. Some of these cutters have two flat face rollers and one cutter roller to prevent raising a burr on the end of the pipe. The flat face rollers iron out the burr and leave the freshly cut pipe the same size clear to the end.

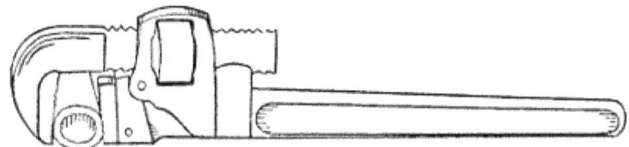

Figure 67.—Pipe-Wrench. This type of wrench is valuable for working with the heavier farm implements. It is intended more for holding than for turning. It is rather rough on nuts. Damaged nuts show signs of careless work.

There should be a good pipe vise that will hold any size pipe up to three inches. At least two pipe wrenches are needed and they should be adjustable from one-quarter-inch up to two-inch pipe.

We must remember that water pipe sizes mean inside measurements. One-inch pipe is about one and one-quarter inches outside diameter. Three-quarter-inch pipe is about one inch outside. Two-inch pipe will carry four times as much water as one-inch pipe, under the rule "doubling the diameter increases the capacity four times."

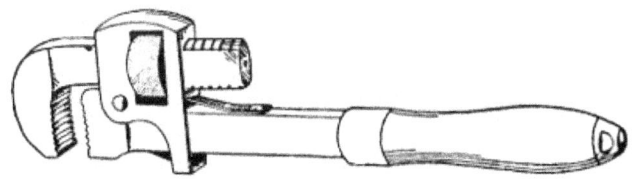

Figure 68.—A smaller sized wrench with wooden handle.

The three-wheel pipe cutter works quickly and is satisfactory for most jobs. Sometimes two of the knife wheels are removed and rollers substituted to prevent raising a burr on the end of the pipe.

Threading dies are made in standard sizes. A good farm set consists of stock and dies to thread all the different sizes of pipe from one-quarter inch to one inch, inclusive. Not many pipes larger than inch are threaded on the farm. They are cut to the proper lengths in the farm shop and the threads are cut in town.

CHAPTER II

FARM SHOP WORK

PROFITABLE HOME REPAIR WORK

Each farmer must be the judge in regard to the kind of mechanical repair work that should be done at home and the kind and amount of repair work that should go to the shop in town. A great deal depends on the mechanical ability of the farmer or his helpers. However, the poorest farm mechanic can do "first aid" service to farm implements and machinery in the nick of time, if he is so disposed. A great many farmers are helpless in this respect because they want to be helpless. It is so much easier to let it go than to go right at it with a determination to fix it, and fix it right.

Figure 69.—Logging Chain. One of the cleverest farm inventions of any age is the logging chain. It is universally used in all lumber camps and on every farm. It usually is from 16 to 20 feet in length, with a round hook on one end for the slip hitch and a grab hook on the other end that makes fast between any two links.

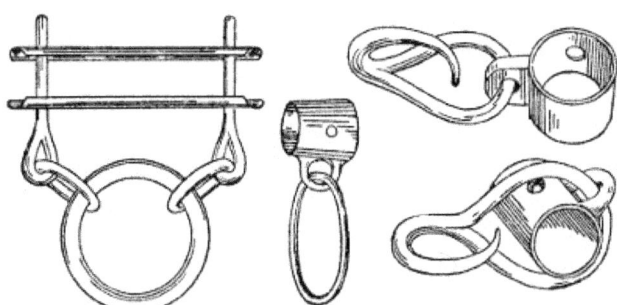

Figure 70.—Neckyoke and Whiffletree Irons. Farmers can make better neckyokes and whiffletrees than they can buy ready-made. The irons may be bought separately and the

wood selected piece by piece.

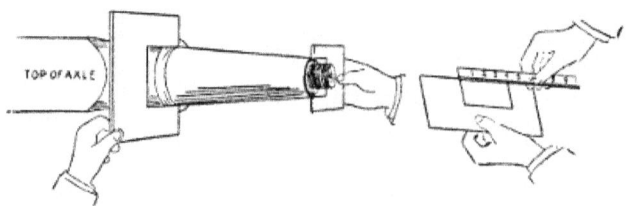

Figure 71.—Measuring a Worn Skein for a New Boxing. The pasteboard calipers are cut to fit the old skein sideways because it is probably flattened on the bottom from wear.

On general principles, however, farm repair work should not occupy a farmer's time to the detriment of growing crops or the proper care of livestock. Farming is the business; mechanical work is a side issue. At the same time, a farmer so inclined can find time during the year to look over every farm machine, every implement and every hand tool on the farm. The stupidest farm helper can clean the rust off of a spade and rub the surface with an oily cloth, in which some fine emery has been dusted. The emery will remove the rust and the oil will prevent it from further rusting. Every laborer knows better than to use a spade or shovel after a rivet head has given way so the handle is not properly supported by the plate extensions. There really is no excuse for using tools or machinery that are out of repair, but the extent to which a farmer can profitably do his own repairing depends on many contingencies. In every case he must decide according to circumstances, always, however, with a desire and determination to run his farm on business principles.

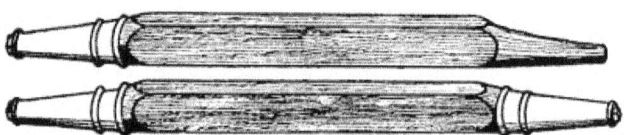

Figure 72.—Wooden Wagon Axles. Axle timber may be bought in the rough or partly fitted to the skeins.

Figure 73.—Showing how to fit the irons on the forward end of a wagon reach.

Figure 74.—Wire Splice. With a little practice wire may be wound close enough to prevent slipping.

Home-made Bolts.—The easiest way to make a bolt is to cut a rod of round iron the proper length and run a thread on each end. On one end the thread may be just long enough to rivet the head, while the thread on the other end is made longer to accommodate the nut and to take up slack. A farmer needs round iron in sizes from one-fourth inch to five-eighths inch. He will use more three-eighths and one-half inch than any other sizes. Blank nuts are made in standard sizes to fit any size of round iron. Have an assortment, in different sizes, of both the square and the hexagon nuts.

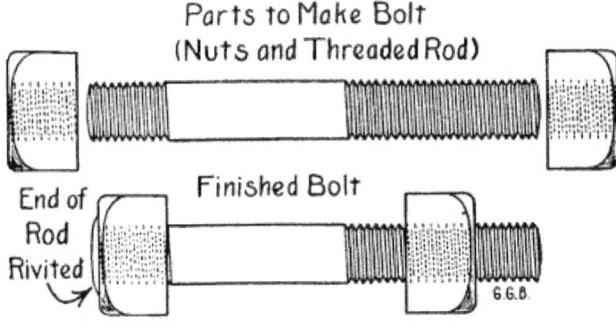

Figure 75.—Emergency Bolts. A bolt may be made quickly without a forge fire by cutting a short thread on one end for the head and a longer thread on the other end for the nut.

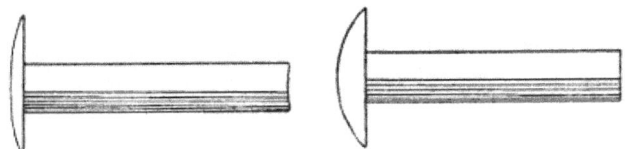

Figure 76.—Rivets. A stock of soft iron rivets of different sizes and lengths should be always kept on hand ready for immediate use.

To make a bolt in the ordinary way requires welding, but for repair work in a hurry it is better to select the proper iron and cut it to the required length either with a cold chisel in the vise, or with a hardy and a handled cold chisel over an anvil. The quickest way of cutting that mashes the rod the least is to be preferred. The size of the rod will determine the manner of cutting in most instances.

Figure 77.—Rivets.

Figure 78.—Rivet Set. This style of set is used for small rivets. The size should be selected to fit the rivets closely. Larger rivets are made to hug the work by means of a flat piece of steel with a hole through it.

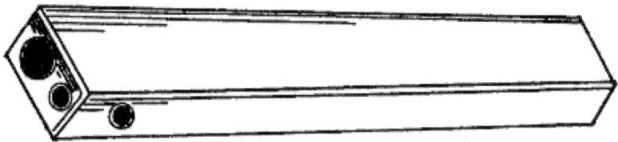

Figure 79.—Rivet Set.

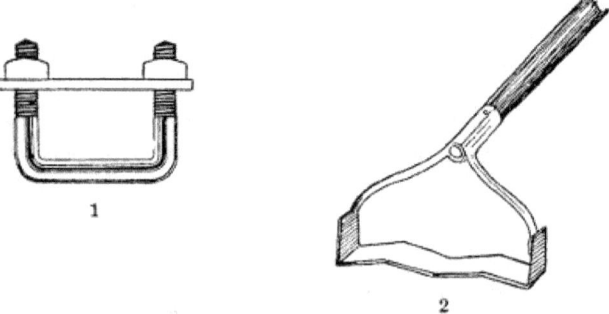

Figure 80.—(1) Coulter Clamp. Plow-beam clamps should be made in the farm shop to fit each plow. (2) Garden Weeder. The quickest hand killer of young weeds in the garden is a flat steel blade that works horizontally half an inch below the surface of the ground.

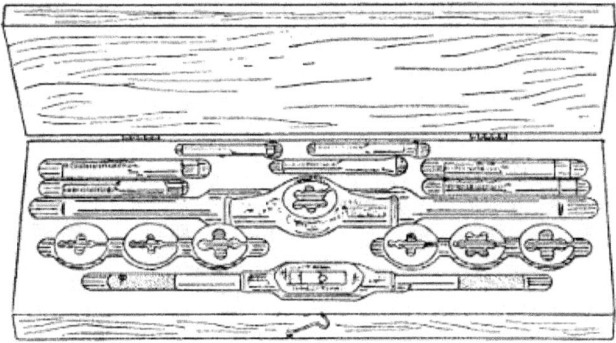

Figure 81.—Stock and Dies. Taps and dies and stocks are best kept in compartments in a case made for the purpose.

Figure 82.—Stock for Round Dies. The opening is turned true and sized accurately to fit. The screw applies pressure to hold the die by friction.

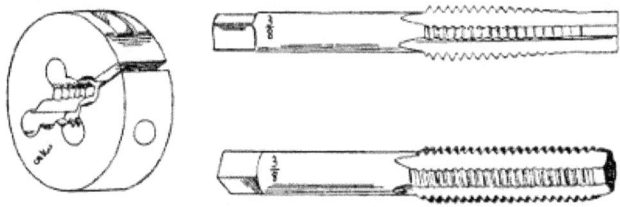

Figure 83.—Taps and Dies. Standard threads are tapped into blank nuts and corresponding threads are cut onto bolts with accuracy and rapidity by using this style taps and dies. They may be had in all sizes. The range for farm work should cut from ¼" to ⅝", inclusive.

Figure 84.—Taper Tap for Blacksmith's Use.

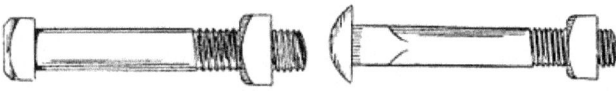

Figure 85.—Machine Bolt and Carriage Bolt. The first is used against iron and the second against wood, but this rule is not arbitrary. The rounded side of the nuts are turned in against wood; the flat side against washers or heavier iron. Use square head bolts if you

expect to take them out after the nuts have rusted on.

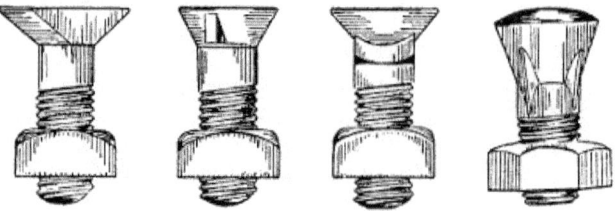

Figure 86.—Plow bolts and sickle bar bolts should be kept in stock. Standard sizes and shapes are made for several different makes of plows and machines.

Taps and dies are made to fit each size of rod. If the thread on the bolt is cut with a solid, or round, plate die, the corresponding tap is run clear through the nut. In that case the nut will screw on the bolt easily, possibly a little loose for some purposes. It is so intended by the manufacturers to give the workman a little leeway. If it is desirable to have the nut screw on the bolt very tight, then the tap is stopped before the last thread enters the nut. A little practice soon qualifies a workman to fit a nut according to the place the bolt is to occupy.

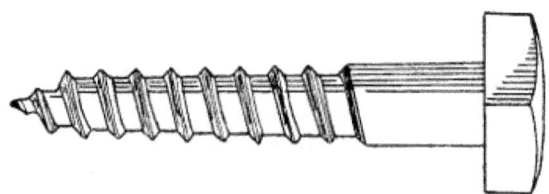

Figure 87.—Lag Screw. To set a lag screw in hardwood, bore a hole the size of the screw shank as calipered between the threads.

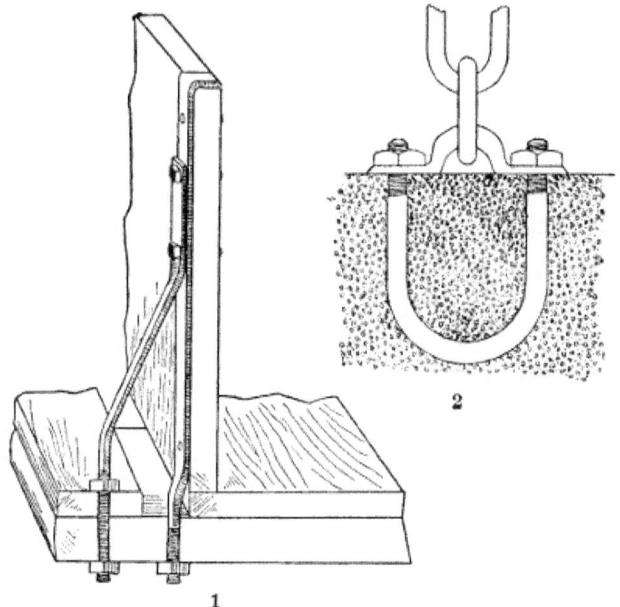

Figure 88.—(1) Wagon-Box Irons, showing how to attach the box and the rave to the cross-piece and to brace the side of the box to hold it upright. There may be several of these braces on each side of the wagon box. (2) U Bolt in Cement. A solid staple to be embedded in concrete for a horse ring, door hinge, cow stanchion, etc.

Generally it is desirable to have nuts fit very snug on parts of machines that shake a good deal, and this applies to almost all farm machinery and implements.

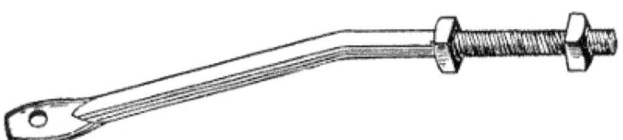

Figure 89.—Wagon-Box Brace. It is offset to hold the rave and to brace the sideboard at the rear and the front ends and sometimes in the middle of light wagon beds.

Figure 90.—Two Plow Clevises and a Plow Link.

Ordinarily a horse rake is supposed to travel steadily along like a cart, but the ground is rough and in practical use the nuts loosen almost as soon as haying commences.

Some farmers make a practice of riveting bolt ends to prevent nuts from working loose. When the bolts have square heads, this practice is not objectionable, because with two wrenches a nut can be twisted off over the riveting, but a great many bolts have round heads and very short, square shanks. Theoretically, the shanks are driven into the wood firm enough to prevent the bolts from turning. Practically this theory is a delusion and a snare, as every farm boy can testify.

Bolts are not manufactured in quantities in the farm blacksmith shop. They can be made by machinery cheaper, but so many times a bolt is needed on short notice that the farm shop should have the necessary tools and materials to supply the need quickly.

Forging Iron and Steel.—Iron and steel are composed of the same properties, but differ chemically. Steel also is finer grained than iron and it requires different treatment. Iron should be forged at a light-red or white heat. If forged at a dark-red heat the iron generally will granulate or crack open and weaken the metal. For a smooth finish the last forging may be done at a dark-red heat, but the hammer must be used lightly. The weight of the hammer as well as the blows also must differ with the different size of iron under heat. Small sizes should be treated with hammer blows that are rather light, while for large sizes the blows should be correspondingly heavy. If light blows be given with a light hammer in forging heavy iron the outside alone will be affected, thus causing uneven tension and contrarywise strain in the iron.

Steel should never be heated above a yellow heat. If heated to a white heat the steel will be burned. Steel should never be forged at a dark-red heat. If this is done it will cause considerable strain between the inner and outer portions, which may cause it to crack while forging. The weight of the hammer and the hammer blows in forging of steel is vastly of more importance than in forging iron. If the blow or the hammer is not heavy enough to exert its force throughout the thickness of the steel it will probably crack in the process of hardening or tempering. If steel be properly forged it will harden easily and naturally, but if improperly forged the tempering will be very difficult—probably a failure. The quality of a finished tool depends greatly upon the correct heat and proper method used in forging and hardening it.

Making Steel Tools.—Steel for tools should first be annealed to even the density and prevent warping. This is done by heating it to a dull cherry red in a slow fire. A charcoal fire for this purpose is best because it contains no sulphur or other injurious impurities. After heating the piece of new steel all over as evenly as possible it should be buried several inches deep in powdered charcoal and left to cool. This completes the annealing process. While working steel into proper shape for tools, great care is required to prevent burning. It should be worked quickly and the process repeated as often as necessary. Practice is the only recipe for speed.

When the tool is shaped as well as possible on the anvil it is then finished with a file by clamping the new tool in the vise, using single cut files. Bastard files are too rough for tool steel. After the tool is shaped by cross-filing and draw-filing to make it smooth it is sometimes polished by wrapping fine emery cloth around the file. Oil is used with emery cloth to give the steel a luster finish. Tempering is the last process in the making of such tools as cold chisels, drills, dies, punches, scratchawls, etc.

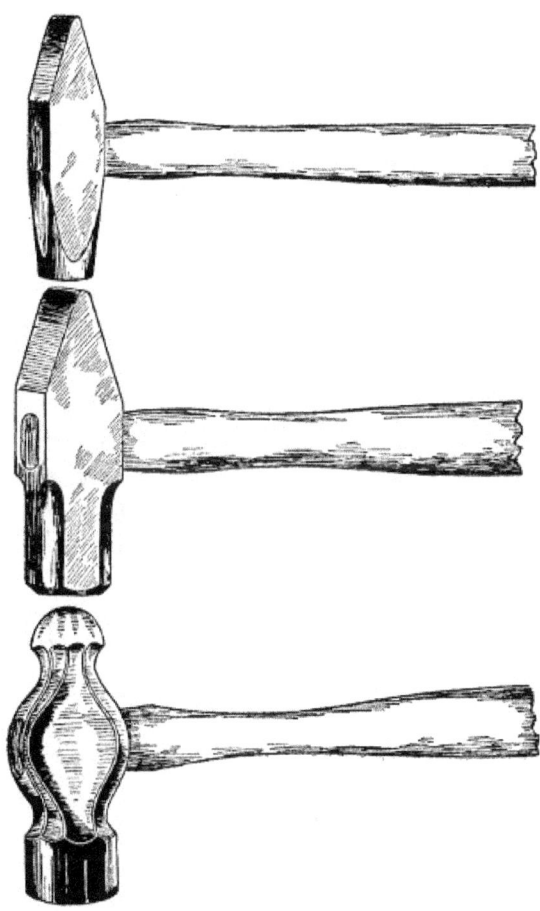

Figure 91.—Blacksmith Hammers. Some smiths use a heavy machinist's hammer. But the flat peen is more useful when working around the anvil and the leg vise.

Tempering Steel Tools.—Good judgment is required to get the right temper. Good eyesight is needed to catch the color at the exact instant, and quick action to plunge it into the water before it cools too much. Dies are made very hard. The color of the steel at dipping time should be a bright straw color. Cold chisels will break when being used if tempered too hard. If cold chisels are to be used for cutting iron, the color should be violet; if the chisels are for cutting stone, purple is the color. Drills for boring iron are tempered a dark straw color at the cutting edge merging back into blue. The water in the dipping tub should be warm, as steel is likely to check or crack when it is tempered in cold water.

Tool steel should be held in a perpendicular position when it enters the water to cool all sides alike. Otherwise the new tool might warp. It is better to dip slowly, sometimes holding the point, or cutting edge, in the water while permitting the shank to cool slowly enough to remain soft. Some sizes of steel may be tempered too hard at first and the temper immediately drawn by permitting the heat of the shank to follow down almost to the edge, then dip. This is done quickly while watching the colors as they move towards the point or edge.

Draw-filing.—Making six-sided and eight-sided punches and scratchawls out of hexagon and octagon tool steel is interesting work. The steel is cut to length by filing a crease all around with a three-cornered file. When it is sufficiently notched, the steel will break straight across. To shape the tool and to draw out the point the steel is heated in the forge to a dull cherry red and hammered carefully to preserve the shape along the taper. Special attention must be given to the numerous corners. A scratchawl or small punch, must be heated many times and hammered quickly before cooling. An old English shop adage reads: "Only one blacksmith ever went to the devil and that was for pounding cold iron."

After the punch or scratchawl is roughed out on the anvil, it is fastened in the vise and finished by cross-filing and draw-filing. Copper caps on the vise jaws will prevent indentations.

Figure 92.—Vise Jaw Guards. Soft auxiliary vise jaws are made of sheet copper or galvanized iron.

Figure 93.—Roll Filing. To file a piece of steel round it is rolled by one hand while the file is used by the other hand.

Draw-filing means grasping each end of the file and moving it back and forth sidewise along the work. For this purpose single-cut files are used. The smoothing is done with a very fine single-cut file, or if very particular, a float file is used. Then the polish is rubbed on with fine emery cloth and oil. The emery cloth is wrapped around the file and the same motion is continued. With some little practice a very creditable piece of work may be turned out. Such work is valuable because of the instruction. A good test of skill at blacksmithing is making an octagon punch that tapers true to the eye when finished.

Set-Screws.—It is customary to fasten a good many gear wheels, cranks and pulleys to machinery shafts by set-screws. There are two kinds of set-screws; one has a cone point, the other a cup end. Both screws are hardened to sink into the shaft. A cup is supposed to cut a ring and the point is supposed to sink into the shaft to make a small hole sufficient to keep the wheel from slipping. However, unless the cone-pointed screw is countersunk into the shaft, it will not hold much of a strain. The point is so small it will slip and cut a groove around the shaft. To prevent this, the set-screw may be countersunk by first marking the shaft with an indentation of the point of the screw. Then the wheel or crank or collar may be removed and a hole drilled into the shaft with a twist-drill the same size, or a sixty-fourth smaller, than the set-screw. Then by forcing the end of the set-screw into the drill hole, the wheel is held solid.

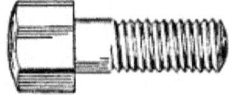 

Figure 94.—Machine-Bolt and Set-Screw. The bolt to the left is used to clamp cylinder heads in place. The set-screw to the right is the cup variety. The end is countersunk to form a cup with a sharp rim.

The principal objection to set-screws is that they are dangerous. The heads always project and are ready to catch a coat sleeve when the shaft is revolving. In all cases, set-screws should be as large as the hub will allow, and it is better to have them protected so it is impossible to catch anything to wind around the shaft. Cup set-screws are not satisfactory except for very light work. If necessary to use them, the ends may be firmly fixed by cutting a ring with a sharp, diamond-point cold chisel.

Setting the Handsaw.—Nine teeth to the inch is the most satisfactory handsaw for all kinds of lumber. Setting the teeth of this kind of saw is best done with a hand lever set. The plunger pin should be carefully adjusted to bend the teeth just far enough to give the necessary set. For general work a saw needs more set than is needed for kiln-dried stuff. The teeth should cut a kerf just wide enough to clear the blade. Anything more is a waste of time and muscle. It is better to work from both sides of the saw by first setting one side the whole length of the blade. Then reverse the saw in the clamp and set the alternate teeth in the same manner. There should be a good solid stop between the handles of the set to insure equal pressure against each sawtooth. The pin should be carefully placed against each tooth at exactly the same spot every time and the pressure should be the same for each tooth.

The best saw-sets for fine tooth saws are automatic so far as it is possible to make them so, but the skill of the operator determines the quality of the work. The reason for setting a saw before jointing is to leave the flattened ends of the teeth square with the blade after the jointing and filing is completed.

Jointing a Handsaw.—After the saw has been set it must be jointed to square the teeth and to even them to equal length, and to keep the saw straight on the cutting edge. Some woodworkers give their saws a slight camber, or belly, to correspond with the sway-back. The camber facilitates cutting to the bottom in mitre-box work without sawing into the bed piece of the box. It also throws the greatest weight of the thrust upon the middle teeth. A saw with even teeth cuts smoother, runs truer and works faster than a saw filed by guess. It is easy to file a saw when all of the teeth are the same length and all have the same set. Anyone can do a good job of filing if the saw is made right to begin with, but no one can put a saw in good working order with a three-cornered file as his only tool.

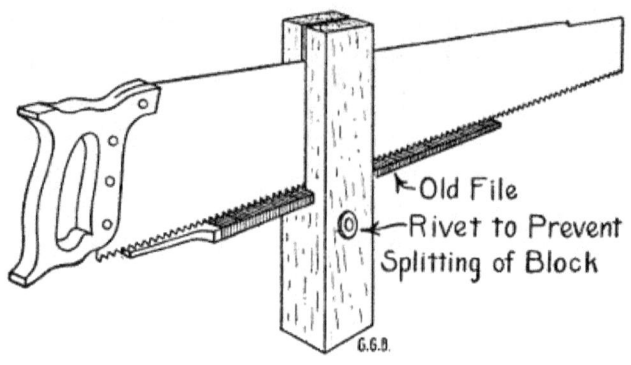

Figure 95.—Saw Jointer. The wooden block is about two inches square by 12″ or 14″ in length. The block is made true and scribed carefully to have the ripsaw slot square, straight and true. The file is set into a mortise square with the block.

Filing the Handsaw.—First comes the three-cornered file. It should be just large enough to do the work. There is no economy in buying larger files thinking that each of the three corners will answer the same purpose as a whole file of smaller size. In the first place the small file is better controlled and will do better work. In the second place the three corners are needed to gum the bottoms of the divisions between the teeth. There is much more wear on the corners than on the sides of a saw-file. Also the corners of a small file are more acute, which means a good deal in the shape of the finished teeth.

After the saw is carefully set and jointed, clamp it in the saw vise and file one side of the saw from heel to point. Then reverse the saw in the saw clamp and file the other side, being careful to keep the bevel of each tooth the same. It is better to stop filing just before the tooth comes to a point. A triangular or diamond shaped point will cut faster and leave a smoother saw kerf and last longer than a needle point.

As the tooth of a crosscut saw is filed away from both edges, it is necessary to make allowances when filing the first side, otherwise some of the teeth will come to a sharp point before the gumming is deep enough.

Using a Handsaw.—Anyone can saw a board square both up and down and crossways by following a few simple rules. Have the board supported on the level by two well made saw-benches 24″ high. Stand up straight as possible and look down on both sides of the saw blade. Use long even strokes and let the saw play lightly and evenly through the saw cut.

Do not cut the mark out; cut to it on the waste end, or further end, if there are more pieces to be cut from the board. The saw kerf is about $3/32$″ wide for a nine-tooth saw set for unkilned lumber or dimension stuff. If both saw kerfs are taken from one piece and none from the next then one length will be ³⁄₁₆″ shorter than the other.

For practice it is a good plan to make two marks ³⁄₃₂″ apart and cut between them. Use a sharp-pointed scratchawl to make the marks. A penknife blade is next best, but it must be held flat against the blade of the square, otherwise it will crowd in or run off at a tangent.

Setting a Circular Saw.—A good saw-set for a circular saw may be made out of an old worn-out flat file. Heat the file in the forge fire to draw the temper and anneal it by covering it with ashes. Smooth it on the grindstone. Put it in the vise and file a notch in one edge. The notch should be just wide enough to fit loosely over the point of a sawtooth. The notch should be just deep enough to reach down one-quarter of the length of the tooth.

Make a saw-set gauge out of a piece of flat iron or steel one inch wide and about four inches long. File a notch into and parallel to one edge at one corner, about one-sixteenth of an inch deep from the edge and about half an inch long measuring from the end. With the home-made saw-set bend the saw teeth outward until the points just miss the iron gauge in the corner notch. The edges of the gauge should be straight and parallel and the notch should be parallel with the edge. In use the edge of the gauge is laid against the side of the saw so the projecting tooth reaches into the notch. One-sixteenth of an inch may be too much set for a small saw but it won't be too much for a 24-inch wood saw working in green cord wood.

Jointing a Circular Saw.—Run the saw at full speed. Lay a 14-inch file flat on the top of the saw table at right angles to the saw. Move the file slowly and carefully towards the saw until it ticks against the teeth. Hold the file firmly by both ends until each sawtooth ticks lightly against the file. A saw in good working order needs very little jointing, but it should have attention every time the saw is set and it should be done after setting and before filing.

Filing a Circular Saw.—The teeth of a crosscut circular saw point a little ahead. Sometimes they point so nearly straight out from the center that you have to look twice to determine which way the saw should run. There are plenty of rules for the pitch of sawteeth, but they are subject to many qualifications. What interests a farmer is a saw that will cut green poles and crooked limbs into stove lengths with the least possible delay. A saw 20 inches in diameter will cut a stick eight inches through without turning it to finish the cut. The front or cutting edges of the teeth of a 24-inch crosscut circular saw for wood sawing should line to a point a little back from the center. This may not sound definite enough for best results, so the more particular farmers may use a straight edge. Select a straight stick about half an inch square. Rest it on top of or against the back of the saw mandrel and shape the forward edges of the teeth on a line with the upper side or rear side of the straight edge. The teeth will stand at the proper pitch when the saw is new, if it was designed for sawing green wood. If it works right before being filed, then the width of the straight edge may be made to conform to the original pitch and kept for future use.

The gumming is done with the edge of the file while filing the front edges of the teeth. It is finished with the flat side of the file while filing the rear edges of the teeth. The depth, or length, of the teeth should be kept the same as the manufacturer designed them. A wood saw works best when the front edges of the teeth have but little bevel. The back edges should have more slant. The teeth should have three-cornered or diamond-shaped points. Needle points break off when they come against knots or cross-grained hardwood. Short teeth do no cutting. Single cut flat files are used for circular saws. The file should fit the saw. It should be about ⅛″ wider than the length of the front side of the teeth. The back edges require that the file shall have some play to show part of the tooth while the file is in motion. Large files are clumsy. The file should be carefully selected.

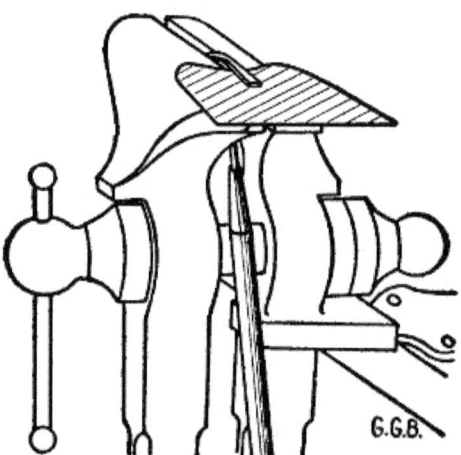

Figure 96.—How to Sharpen a Hoe. Grinding a hoe is difficult, but filing it sharp and straight at the cutting edge is easy. If the hoe chatters when held in the vise, spring a wooden block under the blade. Use false vise jaws to prevent dinging the shank.

How to Sharpen a Hoe.—It is quicker and more satisfactory to file a hoe sharp than to grind it on the grindstone. The shank of the hoe must be held firmly in the vise and there should be a solid block of wood under the blade of the hoe, a little back from the edge; to keep the file from chattering. A single cut flat file is the best to use. It should be long enough to be easily held in one position to make a smooth, even bevel at the same angle to the face of the blade all the way across. To make sure not to file a feather edge it is better to joint the hoe to begin with, then to stop filing just before reaching the edge. If the edge be left $\frac{1}{64}''$ thick it will wear longer and work more easily after having been used an hour or two than it will if the edge be filed thin. This is especially noticeable when the ground contains small stones. Hoes are sharpened from the under side only. The inside of a hoe blade should be straight clear to the edge. Hoes should always have sharp corners. When working around valuable plants you want to know exactly where the corner of the hoe is when the blade is buried out of sight in the ground.

Shoeing Farm Horses.—Farmers have no time or inclination to make a business of shoeing horses, but there are occasions when it is necessary to pull a shoe or set a shoe and to do it quickly. Shoeing tools are not numerous or expensive. They consist first of a tool box, with a stiff iron handle made in the shape of a bale. The box contains a shoeing hammer, hoof rasp, hoof knife, or paring-knife, as it is usually called, and two sizes of horseshoe-nails. Sometimes a foot pedestal is used to set the horse's front foot on when the horse wants to bear down too hard, but this pedestal is not necessary in the farm shop.

There are flat-footed horses that cannot work even in summer without shoes. Common sense and shoeing tools are the only requirements necessary to tack on a plate without calks. Shoes to fit any foot may be purchased at so much a pound.

A paring-knife is used to level the bottom of the hoof so that it will have an even bearing on the shoe all the way round. It is not desirable to pare the frog or the braces in the bottom of a horse's foot. If the foot is well cupped, a little of the horny rim may be taken off near the edges. Generally it is necessary to shorten the toe. This is done partly with the hoof chisel and rasp after the shoe is nailed fast. Sometimes one-fourth of an inch is sufficient; at other times a horse's hoof is very much improved by taking off one-half inch or more of the toe growth either from the bottom or the front or both.

Like all other mechanical work the shoeing of a horse's foot should be studied and planned before starting. A long toe is a bad leverage to overcome when pulling a heavy load. At the same time, nature intended that a horse should have considerable toe length as a protection to the more tender parts of the foot. And the pastern bone should play at the proper angle.

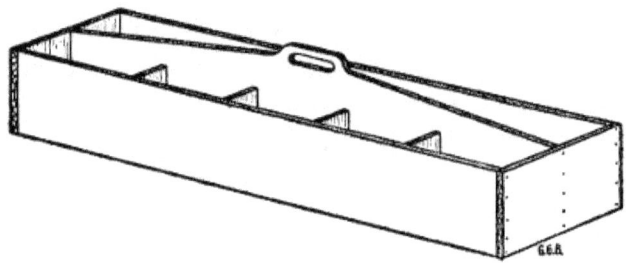

Figure 97.—Tool Box for Field Use. The long open side is for tools. On the other side of the center partition bolts, keys, screws, nails, bits of wire, leather, tin, etc., are kept in the different divisions.

Handy Tool Box.—A tool box with a high lengthwise partition in the middle and a handle in the middle of the top of the partition is the handiest tool box ever used on a farm. At haying and harvest time it should be fitted with the common tools required about haying and harvest machinery. One side is partitioned into square boxes to hold split wire keys, washers, bolts, rivets, and a collection of wire nails, bits of copper wire, a leather punch, etc. On the other side of the box is an assortment of wrenches, cold chisels, punches, pliers and hammers. This tool box belongs in the wagon that accompanies the outfit to the field.

Figure 98.—Melting Ladle. Babbitting shaft boxing requires a melting ladle. It should be about five inches across the bowl and about three inches deep. That is a good size to heat in a forge fire.

Babbitting Boxings.—Babbitting boxings is one of the repair jobs on the farm. Some men are careless about oiling; sometimes sand cuts them out. Every year some boxings need rebabbitting. The melting ladle should be large enough to pour the largest box. Usually a 5-inch bowl is about right. A large ladle will pour a small box but a small ladle won't pour a large one. In cold weather the shaft and box should be warmed to insure an even flow of metal. Pasteboard is fitted against the shaft when pouring the cap or top half of the box. Pasteboard is fitted around the shaft at the ends of the box to keep the melted metal from running out. Never use clay or putty, it is too mussy and the babbitt is made rough and uneven at the edges. Some skill is required to fit either wood or metal close enough to prevent leaks and to do a neat job.

If the boxing is small, both top and bottom may be poured at once by making holes through the dividing pasteboard. The holes must be large enough to let the melted metal through and small enough to break apart easily when cold.

CHAPTER III

GENERATING MECHANICAL POWER TO DRIVE MODERN FARM MACHINERY

At one time ninety-seven per cent of the population of the United States got their living directly from tilling the soil, and the power used was oxen and manual labor. At the present time probably not more than thirty-five per cent of our people are actively engaged in agricultural pursuits. And the power problem has been transferred to horses, steam, gasoline, kerosene and water power, with electricity as a power conveyor.

Fifty years ago a farmer was lucky if he owned a single moldboard cast-iron plow that he could follow all day on foot and turn over one, or at most, two acres. The new traction engines are so powerful that it is possible to plow sixty feet in width, and other machines have been invented to follow the tractor throughout the planting and growing seasons to the end of the harvest. The tractor is supplemented by numerous smaller powers. All of which combine to make it possible for one-third of the people to grow enough to feed the whole American family and to export a surplus to Europe.

At the same time, the standard of living is very much higher than it was when practically everyone worked in the fields to grow and to harvest the food necessary to live.

Farm machinery is expensive, but it is more expensive to do without. Farmers who make the most money are the ones who use the greatest power and the best machinery. Farmers who have a hard time of it are the ones who use the old wheezy hand pump, the eight-foot harrow and the walking plow. The few horses they keep are small and the work worries them. The owner sympathizes with his team and that worries him. Worry is the commonest form of insanity.

Figure 99.—Flail, the oldest threshing machine, still used for threshing pedigreed seeds to prevent mixing. The staff is seven or eight feet long and the swiple is about three feet long by two and one-half inches thick in the middle, tapering to one and one-half inches at the ends. The staff and swiple are fastened together by rawhide thongs.

Figure 100.—Bucket Yoke. It fits around the neck and over the shoulders. Such human yokes have been used for ages to carry two buckets of water, milk or other liquids. The buckets or pails should nearly balance each other. They are steadied by hand to prevent slopping.

At a famous plowing match held at Wheatland, Illinois, two interesting facts were brought out. Boys are not competing for furrow prizes and the walking plow has gone out of fashion. The plowing at the Wheatland plowing match was done by men with riding plows. Only one boy under eighteen years was ready to measure his ability against competition. The attendance of farmers and visitors numbered about three thousand, which shows that general interest in the old-fashioned plowing match is as keen as ever. A jumbo tractor on the grounds proved its ability to draw a big crowd and eighteen plows at the same time. It did its work well and without vulgar ostentation. Lack of sufficient land to keep it busy was the tractor's only disappointment, but it reached out a strong right arm and harrowed the furrows down fine, just to show that it "wasn't mad at nobody."

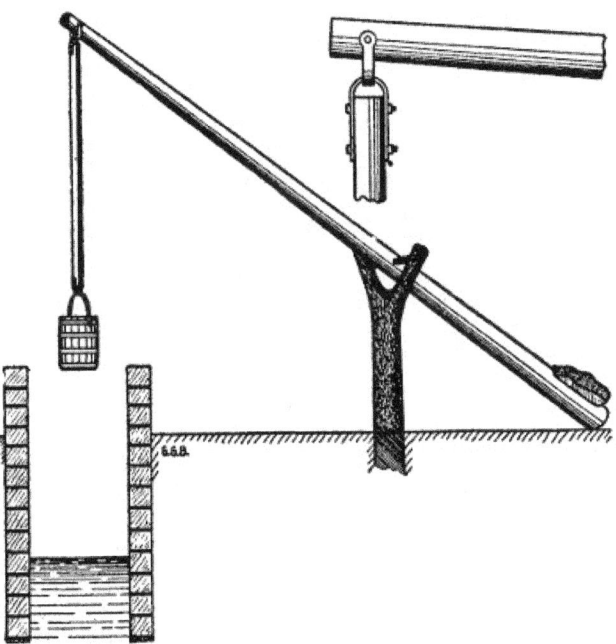

Figure 101.—Well Sweep. The length of the sweep is sufficient to lower the bucket into the water and to raise it to the coping at the top of the brickwork. The rock on the short end of the sweep is just heavy enough to balance the bucket full of water.

Modern farm methods are continually demanding more power. Larger implements are being used and heavier horses are required to pull them. A great deal of farm work is done by engine power. Farm power is profitable when it is employed to its full capacity in manufacturing high-priced products. It may be profitable also in preventing waste by working up cheap materials into valuable by-products. The modern, well-managed farm is a factory and it should be managed along progressive factory methods. In a good dairy stable hay, straw, grains and other feeds are manufactured into high-priced cream and butter.

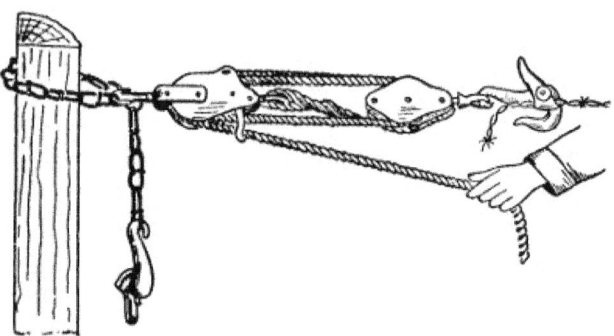

Figure 102.—Wire Stretcher. A small block and tackle will stretch a single barb-wire tight enough for a fence. By using two wire snatches the ends of two wires may be strained together for splicing.

Figure 103.—Block and Tackle. The rope is threaded into two double blocks. There is a safety stop that holds the load at any height.

Farming pays in proportion to the amount of work intelligently applied to this manner of increasing values. It is difficult to make a profit growing and selling grain. Grain may sell for more than the labor and seed, but it takes so much vitality from the land that depreciation of capital often is greater than the margin of apparent profit. When grains are grown and fed to live-stock on the farm, business methods demand better buildings and more power, which means that the farmer is employing auxiliary machinery and other modern methods to enhance values.

In other manufacturing establishments raw material is worked over into commercial products which bring several times the amount of money paid for the raw material.

Figure 104.—Farm Hoists. Two styles of farm elevating hoists are shown in this illustration. Two very different lifting jobs are also shown.

The principle is the same on the farm except that when a farmer raises the raw material he sells it to himself at a profit. When he feeds it to live-stock and sells the live-stock he makes another profit. When the manure is properly handled and returned to the soil he is making another profit on a by-product.

Farming carried on in this way is a complicated business which requires superior knowledge of business methods and principles. In order to conduct the business of farming profitably the labor problem has to be met. Good farm help is expensive. Poor farm help is more expensive. While farm machinery also is expensive, it is cheaper than hand labor when the farmer has sufficient work to justify the outlay. It is tiresome to have agricultural writers ding at us about the superior acre returns of German farms. German hand-made returns may be greater per acre, but one American farmhand, by the use of proper machinery, will produce more food than a whole German family.

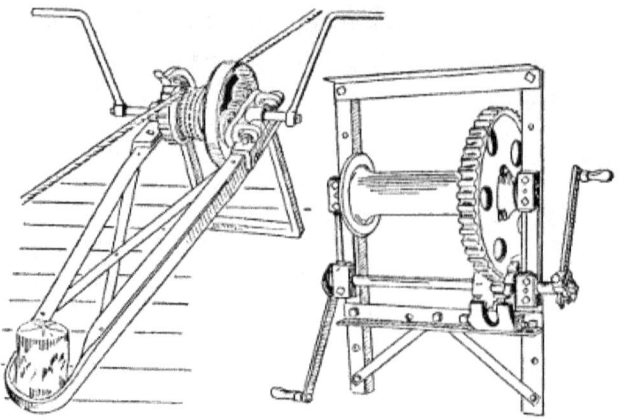

Figure 105.—Two Powerful Winches. The one to the left is used for pulling small stumps or roots in the process of clearing land. The rope runs on and off the drum to maintain three or four laps or turns. The winch to the right is used for hoisting well drilling tools or to hang a beef animal. The rope winds on the drum in two layers if necessary.

DOG CHURN

Even the dog works on some farms. A dog is a nuisance among dairy cattle, but he can be made to earn his salt at churning time. All mechanism in connection with dog power must be light. It also is necessary to eliminate the friction as much as possible.

Figure 106.—Dog Churn Power. A wheel keyed to an iron shaft is placed at an angle as shown. The weight of the dog turns the wheel and power is conveyed to the churn by a light rope belt. It is necessary to confine the dog between stationary partitions built like a stall over the wheel.

The best way to make a dog power is to use a light wooden sulky wheel for the revolving turn table. Next best to the sulky wheel is a light buggy wheel. The wheel is made fast to an upright iron shaft that is stepped into an iron oil well at the bottom and inclined at an angle of about fifteen degrees to give the necessary power. To steady the top of the shaft a light boxing is used, preferably a ball-bearing bicycle race to reduce friction. Power is conveyed to the churn by means of a grooved pulley on the top of the shaft. A small, soft rope or heavy string belt runs from this pulley to a similar pulley connected with the churn.

Dogs learn to like the work when fed immediately after the churning is finished. Dogs have been known to get on to the power wheel to call attention to their hungry condition. This calls to mind the necessity of arranging a brake to stop the wheel to let the dog off. When the wheel is running light, the dog cannot let go.

A spring brake to wear against the iron tire of the wheel is the most satisfactory. The brake may be tripped and set against the tire automatically by a small lever and weight attached to the underside of the wheel. When the speed is too fast the weight swings out and sets the brake. When the speed slackens the weight drops back towards the center and releases the brake. When the speed is about right the weight swings between the two spring catches.

BULL TREADMILL

On dairy farms it is common to see a valuable pure bred bull working a treadmill for exercise and to pump water. Sometimes he turns the cream-separator, but the motion is too unsteady for good results. Treadmills for this purpose are very simple. The mechanism turns a grooved pulley which propels a rope power conveyor. The rope belt may be carried across the yards in any direction and to almost any distance. Bull treadmills consist of a framework of wood which carries an endless apron supported on rollers. The apron link chains pass around and turn two drumhead sprocket-wheels at the upper end and an idler drum at the lower end. The sprocket-wheel drum shaft is geared to an auxiliary shaft which carries a grooved pulley. A rope belt power conveyor runs in this groove and carries power from the bull pen to the pump.

Bull tread powers usually have smooth inclined lags, because a bull's steps on the tread power are naturally uneven and irregular. This construction gives an even straight tread to the travel surface. To prevent slipping, soft wooden strips are nailed onto the lags at the lower edges.

Even incline tread blocks or lags are also recommended for horses that are not shod and for all animals with split hoofs. The traveling apron of the power is placed on an incline and the treads are carried around the two drums at the upper and lower ends of the frame by means of endless chains. There is a governor attachment which regulates the speed and prevents the machinery from "running away."

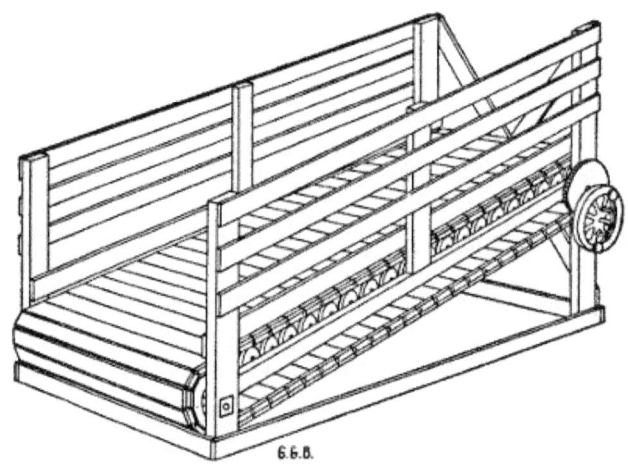

Figure 107.—Bull Tread Power. Treadmills have gone out of fashion. Too much friction was the cause, but a mill like this is valuable to exercise a pure bred bull. Some dairymen make him pump water.

The simplest governor is made on the two-ball governor principle with weights on opposite levers. The governor is attached to two opposite spokes in the flywheel. As the speed increases the weights move outward because of their centrifugal force. This motion operates a brake lever to retard or stop the flywheel. When the machine stops an opposite weight rests against the flywheel until it starts in motion again, so the apron cannot be moved until the brake is released. This is necessary to get the animal on or off of the platform while it is at rest to avoid accidents. The usual incline is a rise of two feet in eight when power is wanted. This pitch compels the bull to lift one-quarter of his own weight and it may be too severe for a heavy animal. The endless apron is an endless hill climb to the bull. Treadmills are not economical of power because there are so many bearings to generate friction.

WINDMILLS

Wind power is the cheapest power we have. A windmill properly proportioned to its work is a great help, especially when it is attached to a good pump for the purpose of lifting water into an elevated tank from which it is piped under pressure for domestic purposes and for watering live-stock.

You can have considerable patience with a windmill if you only depend upon it for pumping water, provided you have a tank that will hold a week's supply to be drawn during a dry, hot time when every animal on the farm demands a double allowance of water. That is the time when a farmer hates to attach himself to the pump handle for the purpose of working up a hickory breeze. That also is the time when the wind neglects a fellow.

A good windmill is useful up to about one-third of its rated capacity, which is the strongest argument for buying a mill larger than at first seems necessary. Some men have suffered at some time in their lives with the delusion that they could tinker with a poorly constructed windmill and make it earn its oil. They have never waked up to a full realization of their early delusion. It is a positive fact that all windmills are not lazy, deceitful nor wholly unreliable. When properly constructed, rightly mounted and kept in good repair, they are not prone to work in a crazy fashion when the tank is full and loaf when it is empty. There are thousands of windmills that have faithfully staid on the job continuously twenty-four hours per day for five or ten years at a stretch, all the time working for nothing year after year without grumbling, except when compelled to run without oil. At such times the protest is loud and nerve racking.

A good windmill with suitable derrick, pump and piping may cost $150. The yearly expense figures something like this:

Interest on investment at 6% per annum	$ 9.00
Depreciation 10%	15.00
Oil	1.00
Repairs	3.00

making a total of $28, which is less than $2.50 per month for the work of elevating a constant supply of water for the house, stable and barnyard.

ONE-MULE PUMP

A home-made device that is much used on live-stock ranches in California is shown in the illustration. This simple mechanism is a practical means for converting circular mule motion into vertical reciprocating pump action. A solid post is set rather deep in the ground about twelve feet from the well. This post is the fulcrum support of the walking-beam. One end of the walking-beam reaches to the center line of the well, where it connects with the pump shaft. The other end of the walking-beam is operated by a pitman shaft connecting with a crank wrist pin near the ground. A round iron shaft similar to a horsepower tumbling rod about ten or twelve feet in length and one and a half inches in diameter is used to convey power and motion to the pitman shaft.

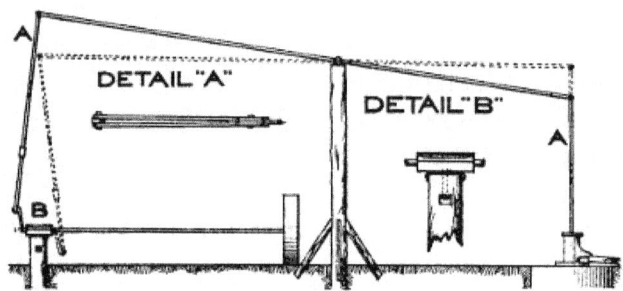

Figure 108.—Mule Pump. A practical home-made power to pump water for live-stock. It is used where the water-table is within 20 feet of the surface of the ground. The drawing shows a post in the center which supports the walking-beam and acts as a fulcrum. A mowing-machine wheel is keyed to one end of a round iron shaft. The other end of this shaft turns in a boxing which is swiveled to a short post as shown at B. See also detail "B.". The two plunger shafts are shown at A A. The mule is hitched to the round iron shaft near the traveling wheel by means of a round hook. As the mule walks around in a circle the shaft revolves and operates the crank B. There are side guys not shown in the drawing to keep the walking-beam in position.

A mowing-machine wheel is keyed to the outer end of the tumbling rod. At the crank end is a babbitted boxing with a bolt attachment reaching down into the top of a short post set solidly into the ground, directly under the inner end of the walking-beam. This bolt permits the boxing to revolve with a swivel motion. Another swivel connects the upper end of the pitman shaft with the walking-beam. The whiffletree is attached to the tumbling rod by an iron hook. This hook is held in place by two iron collars fastened to the tumbling rod by means of keys or set-screws. The mowing-machine drive wheel travels around in a circle behind the mule turning the shaft which works the walking-beam and operates the pump. It would be difficult to design another horse or mule power so cheap and simple and effective. The mule grows wise after a while, so it is necessary to use a blindfold, or he will soldier on the job. With a little encouragement from a whip occasionally a mule will walk around and around for hours pulling the mowing-machine wheel after him.

HORSEPOWER

One horsepower is a force sufficient to lift 33,000 pounds one foot high in one minute.

The term "horsepower" in popular use years ago meant a collection of gear-wheels and long levers with eight or ten horses solemnly marching around in a circle with a man perched on a platform in the center in the capacity of umpire.

This was the old threshing-machine horsepower. It was the first real success in pooling many different farm power units to concentrate the combined effort upon one important operation.

Not many horses are capable of raising 33,000 pounds one foot in one minute every minute for an hour or a day. Some horses are natural-born slackers with sufficient acumen to beat the umpire at his own game. Some horses walk faster than others, also horses vary in size and capacity for work. But during a busy time each horse was counted as one horsepower, and they were only eight or ten in number. And it so developed that the threshing horsepower had limitations which the separator outgrew.

The old threshing horsepower has been superseded by steam engines and gasoline and kerosene power, but horses are more important than ever.

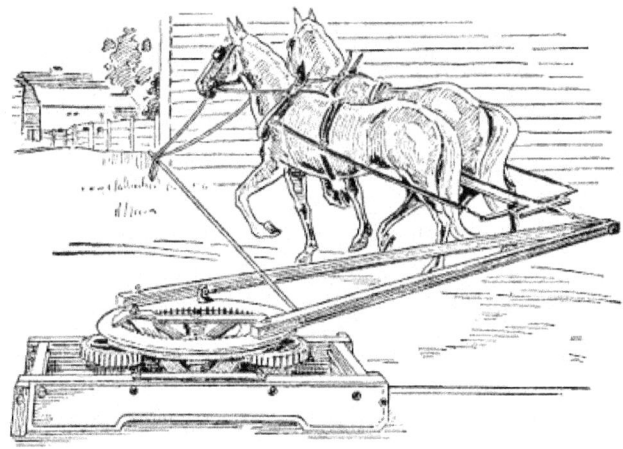

Figure 109.—Horse Power, showing the manner of attaching the braced lever to the bull wheel.

Farm horses are larger and more powerful; they are better kept, better trained, and hitched to better machinery, because it pays. One man drives three 1,600-pound draft horses as fast as he used to drive two 1,000-pound general-purpose horses. The three drafters make play of a heavy load, while the two light horses worry themselves poor and accomplish little. Modern farm machinery is heavier, it cuts wider and digs deeper and does more thorough work. Modern farm requirements go scientifically into the proper cultivation and preparation of soil to increase fertility. Old methods used up fertility until the land refused to produce profitably.

Although the old familiar horsepower has been greatly outclassed, it has not been discarded. There are many small horsepowers in use for elevating grain, baling hay, cutting straw for feed and bedding, grinding feed and other light work where engine power is not available.

WATER-POWER

Water-power is the most satisfactory of all kinds of stationary farm power, when a steady stream of water may be harnessed to a good water-wheel. It is not a difficult engineering feat to throw a dam across a small stream and take the water out into a penstock to supply water to a turbine water-wheel. In the first place it is necessary to measure the flow of water to determine the size of water-wheel which may be used to advantage. In connection with the flow of water it is also important to know the fall. Water is measured by what is termed a "weir." It is easily made by cutting an oblong notch in a plank placed across the stream, as a temporary dam which raises the water a few inches to get a steady, even flow of water through the notch so that calculations may be made in miner's inches. The term "miner's inch" is not accurate, but it comes near enough for practical purposes. Measuring the volume of water should be done during a dry time in summer.

The fall of the stream is easily measured by means of a carpenter's level and a stake. The stake is driven into the ground at a point downstream where water may be delivered to the wheel and a tailrace established to the best advantage. Sighting over the level to a mark on the stake will show the amount of fall. When a manufacturer of water-wheels has the amount of water and the fall, he can estimate the size and character of wheel to supply. The penstock may be vertical or placed on a slant. A galvanized pipe sufficient to carry the necessary amount of water may be laid along the bank, but it should be thoroughly well supported because a pipe full of water is heavy, and settling is likely to break a joint.

Galvanized piping for a farm penstock is not necessarily expensive. It may be made at any tin shop and put together on the ground in sections. The only difficult part about it is soldering the under side of the joints, but generally it may be rolled a little to one side until the bottom of the seam is reached.

The most satisfactory way to carry power from the water-wheel to the farm buildings is by means of electricity. The dynamo may be coupled to the water-wheel and wires carried any required distance.

The work of installing electric power machinery is more a question of detail than mechanics or electrical engineering. The different appliances are bought from the manufacturer and placed where they are needed. It is principally a question of expense and quantity of electricity needed or developed. If the current is used for power, then a motor is connected with the dynamo and current from the dynamo drives the motor. A dynamo may be connected with the water-wheel shaft at the source of power and the motor may be placed in the power-house or any of the other buildings.

The cost of farm waterworks depends principally on the amount of power developed. Small machinery may be had for a few hundred dollars, but large, powerful machinery is expensive. If the stream is large and considerable power is going to waste it might pay to put in a larger plant and sell current to the neighbors for electric lighting and for power purposes. Standard machinery is manufactured for just such plants.

The question of harnessing a stream on your own land when you control both banks is a simple business proposition. If anyone else can set up a plausible plea of riparian rights, flood damage, interstate complications or interference with navigation, it then becomes a question of litigation to be decided by some succeeding generation.

STEAM BOILER AND ENGINE

Farm engines usually are of two different types, steam engines and gasoline or oil engines. Steam stationary engines are used on dairy farms because steam is the best known means of keeping a dairy clean and sanitary. The boiler that furnishes power to run the engine also supplies steam to heat water and steam for sterilizing bottles, cans and other utensils.

For some unaccountable reason steam engines are more reliable than gasoline engines. At the same time they require more attention, that is, the boilers do. Steam engines have been known to perform their tasks year after year without balking and without repairs or attention of any kind except to feed steam and oil into the necessary parts, and occasionally repack the stuffing boxes.

On the other hand, boilers require superintendence to feed them with both fuel and water. The amount of time varies greatly. If the boiler is very much larger than the engine, that is, if the boiler is big enough to furnish steam for two such engines, it will furnish steam for one engine and only half try. This means that the fireman can raise 40 or 60 pounds of steam and attend to his other work around the dairy or barn.

Where steam boilers are required for heating water and furnishing steam to scald cans and wash bottles, the boiler should be several horsepower larger than the engine requirements. There is no objection to this except that a large boiler costs more than a smaller one, and that more steam is generated than is actually required to run the engine. The kind of work required of a boiler and engine must determine the size and general character of the installation.

Portable boilers and engines are not quite so satisfactory as stationary, but there are a great many portable outfits that give good satisfaction, and there is the advantage of moving them to the different parts of the farm when power is required for certain purposes.

SMALL GASOLINE ENGINES

A gasoline engine of 2½ horsepower is the most useful size for a general purpose farm engine. It is convenient to run the pump, washing-machine, fanning-mill, cream-separator, grindstone, and other similar farm chores that have heretofore always been done by human muscle. A small engine may be placed on a low-down truck and moved from one building to another by hand. One drive belt 20 or 30 feet long, making a double belt reach of 12 or 15 feet, will answer for each setting.

The engine once lined up to hitch onto the pulley of any stationary machine is all that is necessary. When the truck is once placed in proper position the wheels may be blocked by a casting of concrete molded into a depression in the ground in front and behind each wheel. These blocks are permanent so that the truck may be pulled to the same spot each time.

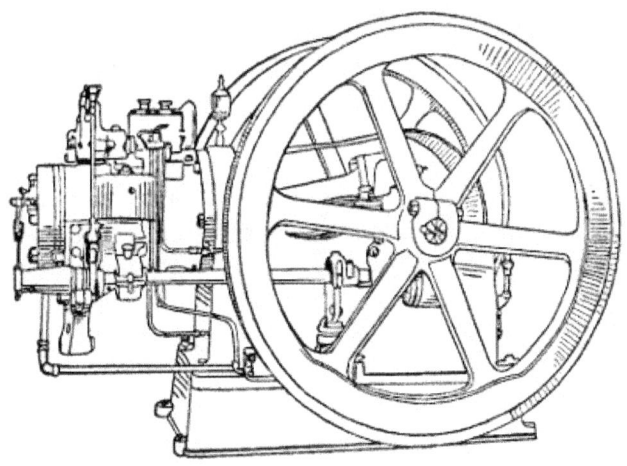

Figure 110.—Kerosene Farm Engine. This is a very compact type of engine with heavy flywheels. A longer base might sit steadier on a wagon, but for stationary use on a solid concrete pier it gives good service.

A gasoline engine for farm use is expected to run by the hour without attention. For this reason it should have a good, reliable hit and miss governor to regulate the speed, as this type is the most economical in fuel. It should have a magneto in addition to a six-cell dry battery. It should be equipped with an impulse starter, a device that eliminates all starting troubles. The engine should be perfectly balanced so as to insure smooth running, which adds materially to the life of the engine. With a good, solid pump jack, a 2½ horsepower engine will pump water until the tank is full, whether it requires one hour or half a day.

It is easily moved to the dairy house to run the separator. As the cream-separator chore comes along regularly every night and every morning, the engine and truck would naturally remain inside of the dairy house more than any other place. If the dairy house is too small to let the engine in, then an addition is necessary, for the engine must be kept under cover. The engine house should have some artistic pretensions and a coat of paint.

KEROSENE PORTABLE ENGINES

The kerosene engine is necessarily of the throttle governor type in order to maintain approximately uniform high temperature at all times, so essential to the proper combustion of kerosene fuel. Therefore, a kerosene engine of the hit-and-miss type should be avoided. However, there are certain classes of work where a throttle governor engine is at a decided disadvantage, such as sawing wood, because a throttle governor engine will not go from light load to full load as quickly as will a hit-and-miss type, and consequently chokes down much easier, causing considerable loss of time.

A general purpose portable kerosene engine is admirably suited to all work requiring considerable horsepower and long hours of service with a fairly steady load, such as tractor work, threshing, custom feed grinding, irrigating and silo filling. There will be a considerable saving in fuel bill over a gasoline engine if the engine will really run with kerosene, or other low-priced fuel, without being mixed with gasoline.

In choosing a kerosene engine, particular attention should be paid to whether or not the engine can be run on all loads without smoking. Unless this can be done, liquid fuel is entering the cylinder which will cause excessive wear on the piston and rings. A good kerosene engine should show as clean an exhaust as when operating on gasoline and should develop approximately as much horsepower. Another feature is harmonizing the fuel oil and the lubricating oil so that one will not counteract the effects of the other.

PORTABLE FARM ENGINE AND TRUCK

A convenient arrangement for truck and portable power for spraying, sawing wood and irrigation pumping, is shown in the accompanying illustration. The truck is low down, which keeps the machinery within reach. The wheels are well braced, which tends to hold the outfit steady when the engine is running. The saw table is detachable. When removed, the spraying tank bolts on to the same truck frame; also the elevated table with the railing around it, where the men stand to spray large apple trees, is bolted onto the wagon bed.

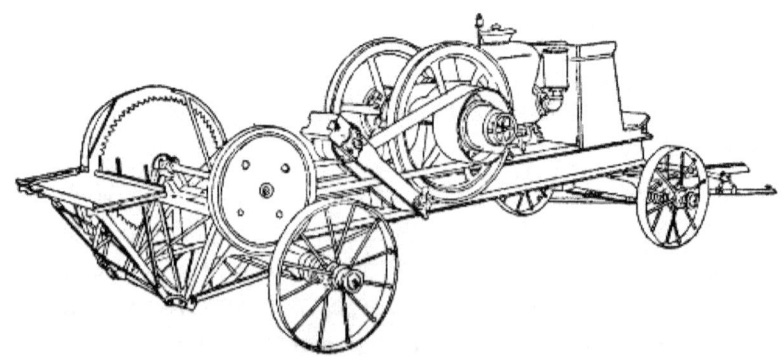

Figure 111.—Portable Farm Engine. This engine is permanently mounted on a low wheel

truck wagon. The saw frame is detachable and the same truck is used for spraying and other work.

Spraying never was properly done until the powerful engine and high pressure tanks were invented. Spraying to be effective, should be fine as mist, which requires a pressure of 150 pounds. There may be a number of attachments to a spraying outfit of this kind. A pipe suspended under the frame with a nozzle for each row is used to spray potatoes, strawberry vines and other low down crops that are grown in rows. When not in use as a portable engine it is blocked firmly into place to run the regular stationary farm machinery.

HYDRAULIC RAM

The hydraulic ram is a machine that gets its power from the momentum of running water. A ram consists of a pipe of large diameter, an air chamber and another pipe of small diameter, all connected by means of valves to encourage the flow of water in two different directions. A supply of running water with a fall of at least two feet is run through a pipe several inches in diameter reaching from above the dam to the hydraulic ram, where part of the flow enters the air chamber of the ram. Near the foot of the large pipe, or at what might be called the tailrace, is a peculiarly constructed valve that closes when running water starts to pass through it. When the large valve closes the water stops suddenly, which causes a back-pressure sufficient to lift a check-valve to admit a certain amount of water from the large supply pipe into the air-chamber of the ram.

After the flow of water is checked, the foot-valve drops of its own weight, which again starts the flow of water through the large pipe, and the process is repeated a thousand or a million times, each time forcing a little water through the check-valve into the air chamber of the ram. The water is continually being forced out into the small delivery pipe in a constant stream because of the steady pressure of the imprisoned air in the air-chamber which acts as a cushion. This imprisoned air compresses after each kick and expands between kicks in a manner intended to force a more or less steady flow of water through the small pipe. The air pressure is maintained by means of a small valve that permits a little air to suck in with the supply of water.

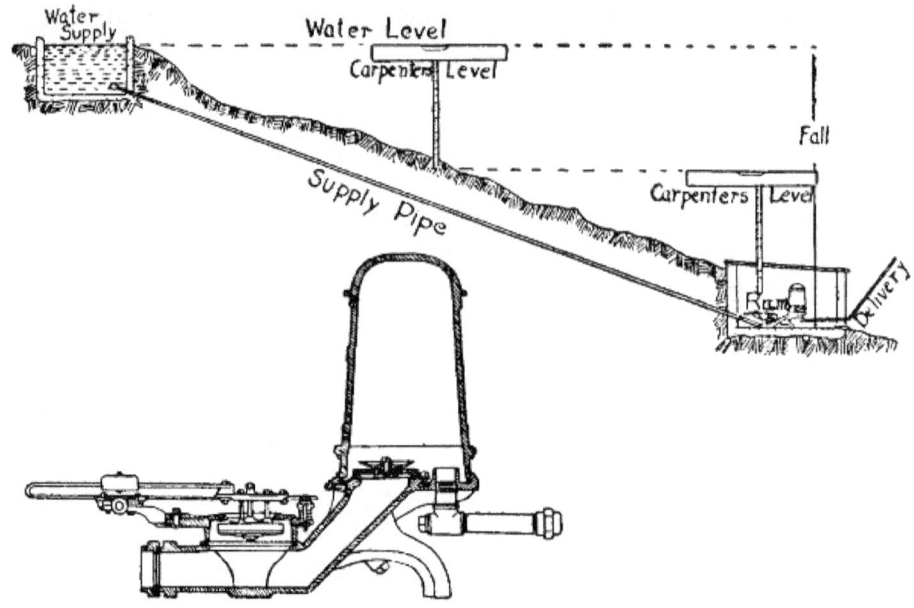

Figure 112.—Hydraulic Ram. The upper drawing shows how to install the ram. The lower drawing is a detail section through the center of the ram. Water flows downhill through the supply pipe. The intermittent action of the valve forces a portion of the water through another valve into the air-chamber. Air pressure forces this water out through delivery pipe. Another valve spills the waste water over into the tailrace. An automatic air-valve intermittently admits air into the air-chamber.

Water may be conveyed uphill to the house by this means, sometimes to considerable distance. The size of the ram and its power to lift water depends upon the amount of water at the spring and the number of feet of fall. In laying the small pipe, it should be placed well down under ground to keep it cool in summer and to bury it beyond the reach of winter frost. At the upper end where the water is delivered a storage tank with an overflow is necessary, so the water can run away when not being drawn for use. A constant supply through a ram demands a constant delivery. It is necessary to guard the water intake at the dam. A fence protection around the supply pool to keep live-stock or wild animals out is the first measure of precaution. A fine screen surrounding the upper end of the pipe that supplies water to the ram is necessary to keep small trash from interfering with the valves.

THE FARM TRACTOR

Farm tractors are becoming practical. Most theories have had a try out, the junk pile has received many failures and the fittest are about to survive.

Now, if the manufacturers will standardize the rating and the important parts and improve their selling organizations the whole nation will profit. The successful tractors usually have vertical engines with four cylinders. They are likely to have straight spur transmission gears, and a straight spur or chain drive, all carefully protected from dust. And they will have considerable surface bearing to avoid packing the soil. Some tractors carry their weight mostly upon the drive wheels—a principle that utilizes weight to increase traction. Other tractors exert a great deal of energy in forcing a small, narrow front steering-wheel through the soft ground. Any farmer who has pushed a loaded wheelbarrow knows what that means. Some kerosene tractors require a large percentage of gasoline. The driver may be as much to blame as the engine. But it should be corrected.

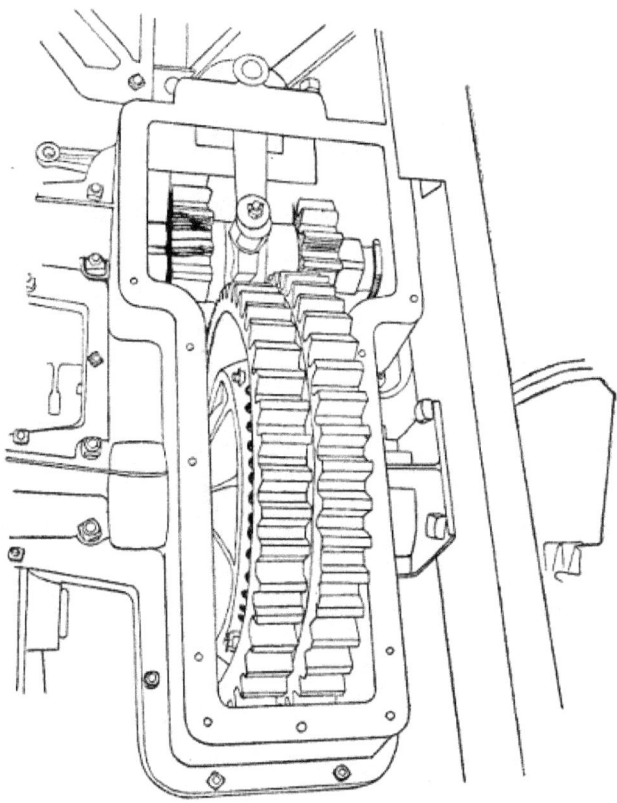

Figure 113.—Tractor Transmission Gear. Spur gears are the most satisfactory for heavy work.

Manufacturers should do more educational work and talk less about the wonderfully marvelous and marvelously wonderful. Salesmen should study mechanics instead of oratory. Tractor efficiency should be rated practically instead of theoretically. The few actual reports of performance have emanated from tests with new machines in the hands of trained demonstrators. Manufacturers include belt power work among the virtues of farm tractors, and they enumerate many light jobs, such as running a cream-separator, sawing wood, pumping water and turning the fanning-mill. Well, a farm tractor can do such work—yes. So can an elephant push a baby carriage. If manufacturers would devise a practical means of using electricity as an intermediary, and explain to farmers how a day's energy may be stored in practical working batteries to be paid out in a week, then we could understand why we should run a 20 horsepower engine to operate a cream-separator one hour at night and another hour in the morning.

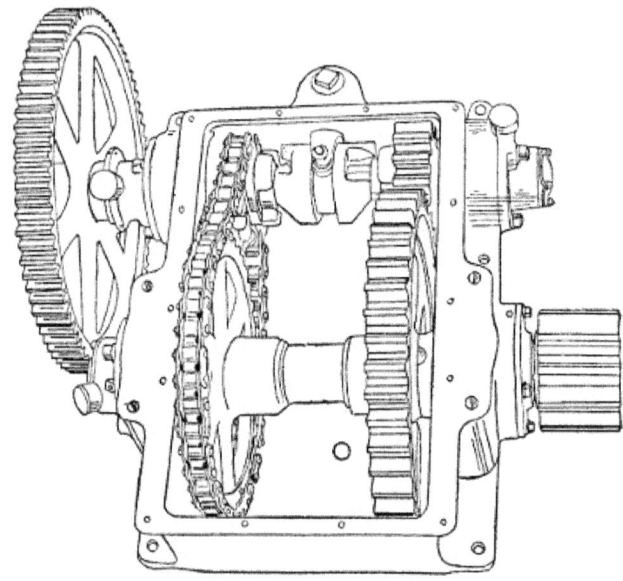

Figure 114.—Straight Transmission Gear, forward and chain drive reverse, for traction engine.

CHAPTER IV

DRIVEN MACHINES

FARM WATERWORKS

Every farm has its own water supply. Some are very simple, others are quite elaborate. It is both possible and practical for a farmer to have his own tap water under pressure on the same plan as the city. When good water is abundant within 75 feet of the surface of the ground the farm supply may be had cheaper and better than the city. Even deep well pumping is practical with good machinery rightly installed. Farm waterworks should serve the house and the watering troughs under a pressure of at least 40 pounds at the ground level. The system should also include water for sprinkling the lawn and for irrigating the garden. If strawberries or other intensive money crops are grown for market there should be sufficient water in the pipes to save the crop in time of drouth. These different uses should all be credited to the farm waterworks system pro rata, according to the amounts used by the different departments of the farm. The books would then prove that the luxury of hot and cold running water in the farmhouse costs less than the average city family pays.

Three Systems of Water Storage.—The first plan adopted for supplying water under pressure on farms was the overhead tank. The water was lifted up into the tank by a windmill and force pump. Because wind power proved rather uncertain farmers adopted the gasoline engine, usually a two horsepower engine.

The second water storage plan was the air-tight steel water-tank to be placed in the cellar or in a pit underground. The same pump and power supplies the water for this system, but it also requires an air-pump to supply pressure to force the water out of the tank.

The third plan forces the water out of the well by air pressure, as it is needed for use. No water pump is required in this system; the air-compressor takes its place.

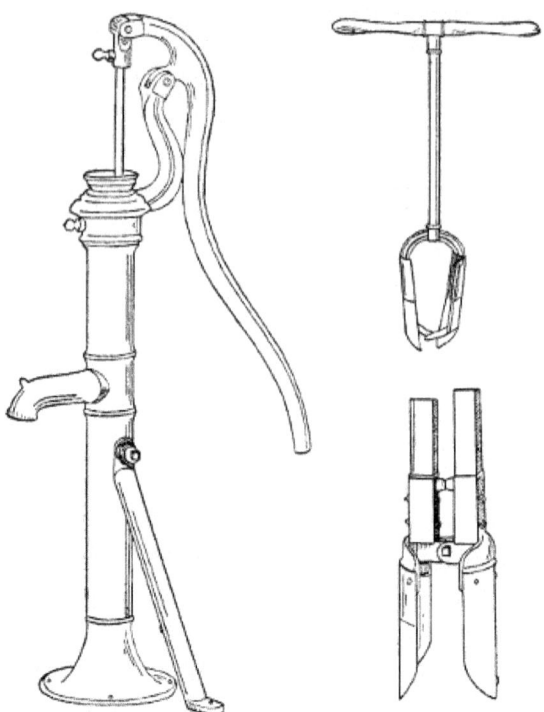

Figure 115.—The Farm Pump. It superseded the iron-bound bucket, the slimy old bucket, the malaria-lined bucket that hung in the well, but it wore out the women. Oil was never wasted on its creaking joints. Later it was fitted with a stuffing-box and an air-chamber, and the plunger was hitched to the windmill.

To the right are shown two kinds of post-hole diggers. The upper digger is sometimes used to clear the fine earth out of the bottom of a hole dug by the lower digger.

Suction-Pumps.—The word suction, when applied to pumps, is a misnomer. The principle upon which such pumps work is this: The pump piston drives the air out of the pump cylinder which produces a vacuum. The pressure of the atmosphere is about fifteen pounds per square inch of surface. This pressure forces sufficient water up through the so-called suction pipe to fill the vacuum in the cylinder. The water is held in the cylinder by foot-valves or clack-valves. As the piston again descends into the cylinder it plunges into water instead of air. A foot-valve in the bottom end of the hollow piston opens while going down and closes to hold and lift the water as the piston rises. Water from the well is forced by atmospheric pressure to follow the piston and the pump continues to lift water so long as the joints remain air-tight. The size of piston and length of stroke depend on the volume of water required, the height to which it must be lifted and the power available. A small power and a small cylinder will lift a small quantity of water to a considerable height. But increasing the volume of water requires a larger pump and a great increase in the power to operate it. The size of the delivery pipe has a good deal to do with the flow of water. When water is forced through a small pipe at considerable velocity, there is a good deal of friction. Often the amount of water delivered is reduced because the discharge pipe is too small. Doubling the diameter of a pipe increases its capacity four times. Square turns in the discharge pipe are obstructions; either the pipe must be larger or there will be a diminished flow of water. Some pump makers are particular to furnish easy round bends instead of the ordinary right-angled elbows. A great many pumps are working under unnecessary handicaps, simply because either the supply pipe or discharge pipe is not in proportion to the capacity of the pump, or the arrangement of the pipes is faulty.

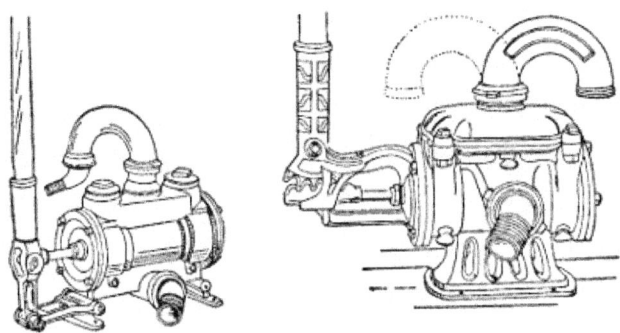

Figure 116.—Hand Force-Pump. Showing two ways of attaching wooden handles to hand force-pumps.

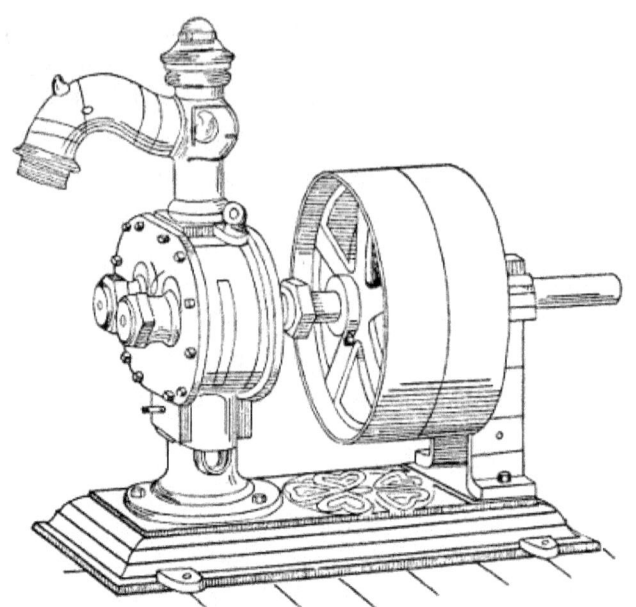

Figure 117.—Rotary Pump. Twin water-chamber rotary pumps take water through the bottom and divide the supply, carrying half of the stream around to the left and the other half to the right. The two streams meet and are discharged at the top.

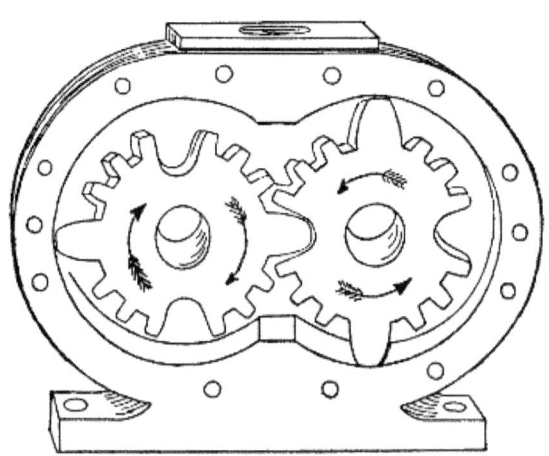

Figure 118.—Section of Rotary Pump.

Rotary Pumps.—A twin-chamber rotary pump admits water at the bottom of the chamber and forces it out through the top. Intermeshing cogs and rotary cams revolve outward from the center at the bottom, as shown by the arrows in Figure 118. The stream of water is divided by the cams, as it enters the supply pipe at the bottom, and half of the water is carried each way around the outsides of the double chamber. These streams of water

78

meet at the top of the chamber, where they unite to fill the discharge pipe. These pumps operate without air-chambers and supply water in a continuous stream. They may be speeded up to throw water under high pressure for fire fighting, but for economy in ordinary use the speed is kept down to 200 revolutions, or thereabout. Rotary pumps are also made with one single water chamber cylinder. The pump head, or shaft, is placed a little off center. A double end cam moves the water. Both ends of the cam fit against the bore of the cylinder. It works loosely back and forth through a slotted opening in the pump head. As the shaft revolves the eccentric motion of the double cam changes the sizes of the water-pockets. The pockets are largest at the intake and smallest at the discharge. Rotary pumps are comparatively cheap, as regards first cost, but they are not economical of power. In places where the water-table is near the surface of the ground they will throw water in a very satisfactory manner. But they are more used in refineries and factories for special work, such as pumping oil and other heavy liquids.

Centrifugal Pumps.—The invention and improvement of modern centrifugal pumps has made the lifting of water in large quantities possible. These pumps are constructed on the turbine principle. Water is lifted in a continuous stream by a turbine wheel revolving under high speed. Water is admitted at the center and discharged at the outside of the casing. Centrifugal pumps work best at depths ranging from twenty to sixty feet. Manufacturers claim that farmers can afford to lift irrigation water sixty feet with a centrifugal pump driven by a kerosene engine.

The illustrations show the principle upon which the pump works and the most approved way of setting pumps and engines. Centrifugal pumps usually are set in dry wells a few feet above the water-table. While these pumps have a certain amount of suction, it is found that short supply pipes are much more efficient. Where water is found in abundance within from 15 to 30 feet of the surface, and the wells may be so constructed that the pull-down, or the lowering of the water while pumping is not excessive, then it is possible to lift water profitably to irrigate crops in the humid sections. Irrigation in such cases, in the East, is more in the nature of insurance against drouth. Valuable crops, such as potatoes and strawberries, may be made to yield double, or better, by supplying plenty of moisture at the critical time in crop development. It is a new proposition in eastern farming that is likely to develop in the near future.

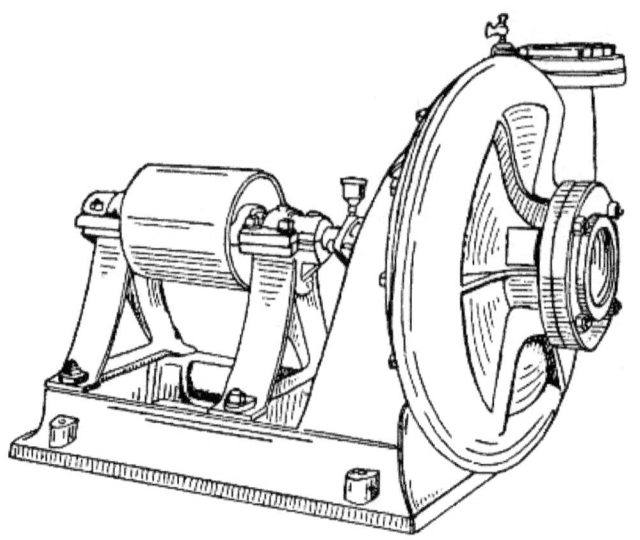

Figure 119.—Centrifugal Pump. This style of pump is used in many places for irrigation. It runs at high speed, which varies according to the size of the pump. It takes water at the center and discharges it at the outside of the casing.

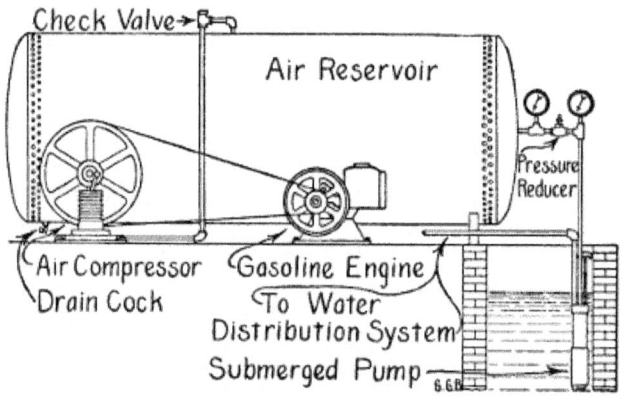

Figure 120.—Air Pressure Pump. Pumping water by air pressure requires a large air container capable of resisting a pressure of 100 pounds per square inch. This illustration shows the pressure tank, engine, air-compressor, well and submerged pump.

Air Pressure Pump.—Instead of pumping water out of the well some farmers pump air into the well to force the water out. A double compartment cylindrical tank is placed in the water in the well. These tanks are connected with the farm water distributing system to be carried in pipes to the house and to the stock stables. Air under a pressure of from 50 to 100 pounds per square inch is stored in a steel tank above ground. Small gas-pipes connect this air pressure tank with the air-chamber of the air-water tank in the well. A peculiar automatic valve regulates the air so that it enters the compartment that is filled, or partly filled, with water, and escapes from the empty one so the two compartments work together alternately. That is, the second chamber fills with water, while the first chamber is being drawn upon. Then the first chamber fills while the second is being emptied. This system will work in a well as small as eight inches in diameter, and to a depth of 140 feet. It might be made to work at a greater depth, but it seems hardly practical to do so for the reason that, after allowing for friction in the pipes, 100 pounds of air pressure is necessary to lift water 150 feet. An air tank of considerable size is needed to provide storage for sufficient air to operate the system without attention for several days. Careful engineering figures are necessary to account for the different depths of farm wells, and the various amounts of water and power required. For instance: The air tank already contains 1,000 gallons of air at atmospheric pressure—then: Forcing 1,000 gallons of atmospheric air into a 1,000-gallon tank will give a working pressure of 15 pounds per square inch; 2,000 gallons, 30 pounds; 3,000 gallons, 45 pounds, and so on. Therefore, a pressure of 100 pounds in a 1,000-gallon tank (42 inches by 14 feet) would require 6,600 gallons of free atmosphere, in addition to the original 1,000 gallons, and the tank would then contain 1,000 gallons of compressed air under a working pressure of 100 pounds per square inch. A one cylinder compressor 6 inches by 6 inches, operating at a speed of 200 R.P.M. would fill this tank to a working pressure of 100 pounds in about 50 minutes. One gallon of air will deliver one gallon of water at the faucet. But the air must have the same pressure as the water, and there must be no friction. Thus, one gallon of air under a working pressure of forty-five pounds, will, theoretically, deliver one gallon of water to a height of 100 feet. But it takes three gallons of free air to make one gallon of compressed air at forty-five pounds pressure. If the lift is 100 feet, then 1,000 gallons of air under a pressure of forty-five pounds will theoretically deliver 1,000 gallons of water. Practically, the air tank would have to be loaded to a very much greater pressure to secure the 1,000 gallons of water before losing the elasticity of the compressed air. If one thousand gallons of water is needed on the farm every day, then the air pump would have to work about one hour each morning. This may not be less expensive than pumping the water directly, but it offers the advantage of water fresh from the well. Pure air pumped into the well tends to keep the water from becoming stale.

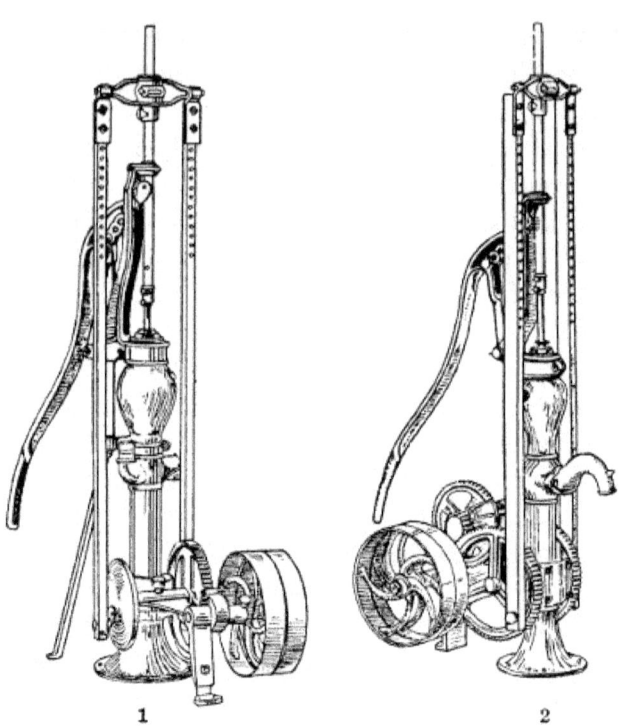

Figure 121.—(1) Single-Gear Pump Jack. This type of jack is used for wells from 20 to 40 feet deep. (2) Double-Gear, or Multiple-Gear Pump Jack. This is a rather powerful jack designed for deep wells or for elevating water into a high water-tank.

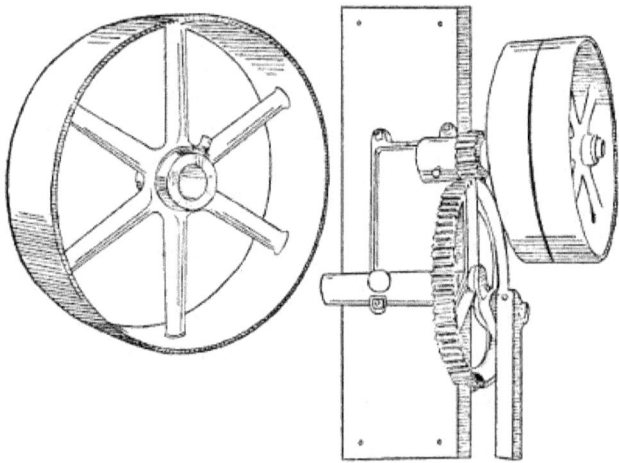

Figure 122.—Post Pump Jack. This arrangement is used in factories when floor space is valuable. The wide-face driving-pulley is shown to the left.

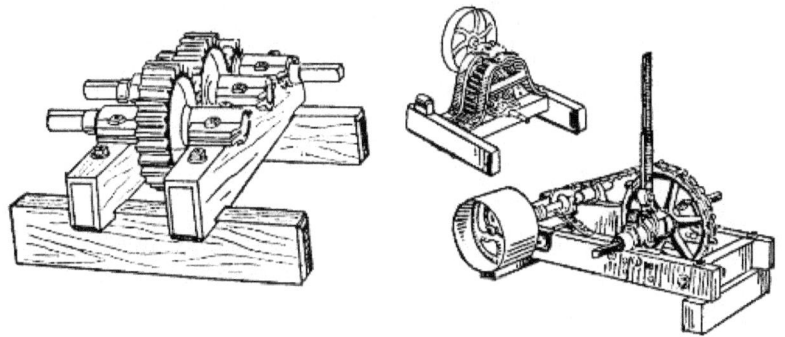

Figure 123.—Three Jacks for Different Purposes. At the left is a reverse motion jack having the same speed turning either right or left. The little jack in the center is for light work at high belt speed. To the right is a powerful jack intended for slow speeds such as hoisting or elevating grain.

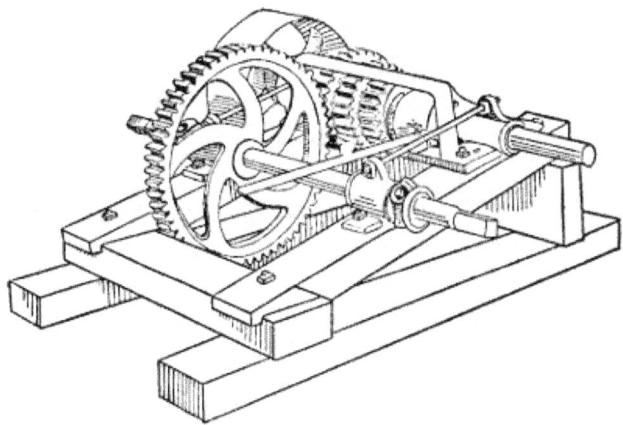

Figure 124.—Speed Jack, for reducing speed between engine and tumbling rod or to increase speed between tumbling rod and the driven machine.

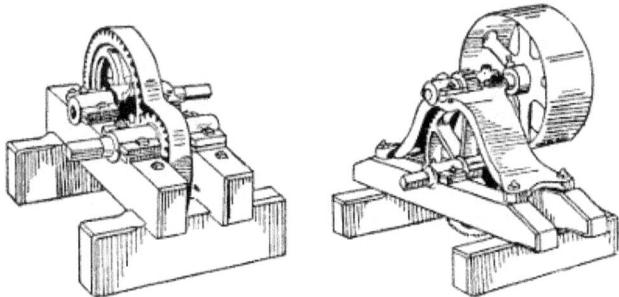

Figure 125.—The Speed Jack on the left is used either to reduce or increase tumbling rod speed and to reverse the motion. The Speed Jack on the right transfers power either from belt to tumbling rod or reverse. It transforms high belt speed to low tumbling rod speed, or vice versa.

Pump Jacks and Speed Jacks.—Farm pumps and speed-reducing jacks are partners in farm pumping. Force-pumps should not run faster than forty strokes per minute. Considerable power is required to move the piston when the water is drawn from a deep well and forced into an overhead tank. Jacks are manufactured which bolt directly to the pump, and there are pumps and jacks built together. A pump jack should have good, solid gearing to reduce the speed. Spur-gearing is the most satisfactory. Bevel-gears are wasteful of power when worked under heavy loads. Power to drive a pump jack is applied to a pulley at least twelve inches in diameter with a four-inch face when belting is used. If a rope power conveyor is used, then pulleys of larger diameters are required to convey the same amount of power.

Only general terms may be used in describing the farm pump, because the conditions differ in each case. Generally speaking, farmers fail to appreciate the amount of power used, and they are more than likely to buy a jack that is too light. Light machinery may do the work, but it goes to pieces quicker, while a heavy jack with solid connections will operate the pump year in and year out without making trouble. For increasing or reducing either speed or power some kind of jack is needed. All farm machines have their best speed. A certain number of revolutions per minute will accomplish more and do better work than any other speed. To apply power to advantage speed jacks have been invented to adjust the inaccuracies between driver and driven.

IRRIGATION BY PUMPING

The annual rainfall in the United States varies in different parts of the country from a few inches to a few feet. Under natural conditions some soils get too much moisture and some too little. Irrigation is employed to supply the deficiency and drainage, either natural or artificial, carries off the excess. Irrigation and drainage belong together. Irrigation fills the soil with moisture and drainage empties it. Thus, a condition is established that supplies valuable farm plants with both air and moisture. In the drier portions of the United States, nothing of value will grow without irrigation. In the so-called humid districts deficiency of moisture at the critical time reduces the yield and destroys the profit. The value of irrigation has been demonstrated in the West, and the practice is working eastward.

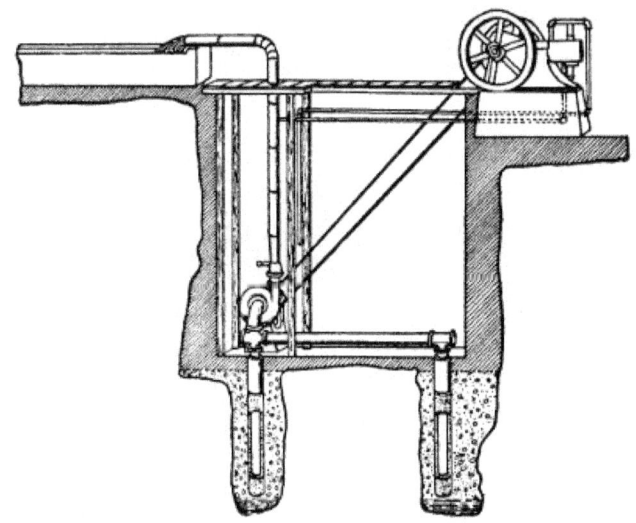

Figure 126.—Centrifugal Pump Setting. When used for irrigation, centrifugal pumps are set as close to the ground water as practical.

Irrigation is the new handmaiden of prosperity. A rainy season is a bountiful one. Irrigation supplies the bounty without encouraging destructive fungus diseases. Where water is abundant within easy reach, pumping irrigation water is thoroughly practical. Improvements in pumps in recent years have increased their capacity and insured much greater reliability. A centrifugal pump is recommended for depths down to 75 feet; beyond this depth the necessity of installing more expensive machinery places the business of pumping for irrigation on a different plane. A centrifugal pump will throw more water with less machinery than any other device, but like all other mechanical inventions, it has its limitations. In figuring economical pumping, the minimum quantity should be at least 100 gallons per minute, because time is an object, and irrigation, if done at all, should cover an area sufficient to bring substantial returns. Centrifugal pumps should be placed near the surface of the water in the well. For this reason, a large, dry well is dug down to the level of the water-table and the pump is solidly bolted to a concrete foundation built on the bottom of this well. A supply pipe may be extended any depth below the pump, but the standing water surface in the well should reach within a few feet of the pump. The pump and supply must be so well balanced against each other that the pull-down from pumping will not lower the water-level in the well more than twenty feet below the pump. The nearer the ground water is to the pump the better.

The water well below the pump may be bored, or a perforated well pipe may be driven; or several well points may be connected. The kind of well must depend upon the condition of the earth and the nature of the water supply. Driven wells are more successful when water is found in a stratum of coarse gravel.

Before buying irrigation machinery, it is a good plan to test the water supply by temporary means. Any good farm pump may be hitched to a gasoline engine to determine if the water supply is lasting or not. Permanent pumping machinery should deliver the water on high ground. A main irrigation ditch may be run across the upper end of the field. This ditch should hold the water high enough so it may be tapped at convenient places to run through the corrugations to reach the roots of the plants to be benefited. There are different systems of irrigation designed to fit different soils. Corrugations are the cheapest and the most satisfactory when soils are loose enough to permit the water to soak into the soil sideways, as well as to sink down. The water should penetrate the soil on both sides of the corrugations for distances of several inches. Corrugations should be straight and true and just far enough apart so the irrigation water will soak across and meet between. Some soils will wash or gully out if the fall is too rapid. In such cases it may be necessary to terrace the land by following the natural contour around the ridges so the water may flow gently. Where the fall is very slight, that is, where the ground is so nearly level that it slopes away less than six inches in a hundred feet, it becomes necessary to prepare the land by building checks and borders to confine the water for a certain length of time. Then it is let out into the next check. In the check and border system the check bank on the lower side has an opening which is closed during the soaking period with a canvas dam. When the canvas is lifted the water flows through and fills the next check. This system is more expensive, and it requires more knowledge of irrigation to get it started, and it is not likely to prove satisfactory in the East.

For fruits and vegetables, what is known as the furrow system of irrigation is the most practical. An orchard is irrigated by plowing furrows on each side of each row of trees. The water is turned into these furrows and it runs across the orchard like so many little rivulets. Potatoes are irrigated on the same plan by running water through between the rows after the potatoes have been ridged by a double shovel-plow. This plan also works well with strawberries. After the land is prepared for irrigation, the expense of supplying water to a fruit orchard, strawberry patch or potato field is very little compared with the increase in yield. In fact, there are seasons when one irrigation will save the crop and produce an abundant yield, when otherwise it would have been almost a total loss.

Overhead Spray Irrigation.—The most satisfactory garden irrigation is the overhead spray system. Posts are set ten feet apart in rows 50 feet apart. Water pipes are laid on the tops of the posts and held loosely in position by large staples. These water pipes are perforated by drilling a line of small holes about three feet apart in a straight line along one side of the pipe. The holes are tapped and small brass nozzles are screwed in. The overhead pipes are connected with standpipes at the highest place, generally at the ends of the rows. The pipe-lines are loosely coupled to the standpipes to permit them to roll partly around to direct the hundreds of spray nozzles as needed.

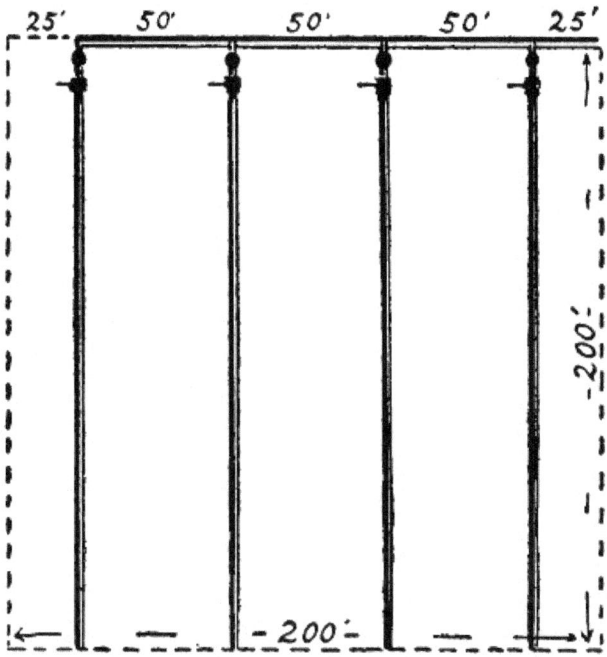

Figure 127.—Overhead Irrigation. Diagram showing the arrangement of pipes for irrigating one acre of land. The pipes are supported on posts six feet high.

Six feet high is sufficient to throw a fine mist or spray twenty-five feet, which is far enough to meet the spray from the next row, so the ground will be completely covered. To do this the pipes are rolled from one side to the other, through a 90 degree arc to throw the spray on both sides. The pipes usually are laid with a grade which follows down the slope of the land. A fall of one foot in fifty is sufficient. Water is always admitted at the upper end of each pipe-line to flow down by gravity, assisted by tank pressure. A pressure of about forty pounds is needed to produce a fine spray, and to send it across to meet the opposite jets. The little brass nozzles are drilled with about a one-eighth inch hollow. But the jet opening is small, about No. 20 W. G. This gives a wire-drawn stream that quickly vaporizes when it meets the resistance of the atmosphere. When properly installed a fine misty rain is created, which quickly takes the same temperature as the air, and settles so gently that the most delicate plants are not injured.

Quantity of Water to Use.—Good judgment is necessary in applying water to crops in regard to quantity, as well as the time of making application. Generally speaking, it is better to wait until the crop really needs moisture. When the pump is started give the crop plenty with the expectation that one irrigation will be sufficient. Much depends upon the amount of moisture in the soil; also the kind of crop and weather conditions enter into the problem. On sandy land that is very dry where drainage is good, water may be permitted to run in the corrugations for several days until the ground is thoroughly soaked. When potatoes are forming, or clover is putting down its big root system, a great deal of water is needed. Irrigation sufficient to make two inches of rainfall may be used to advantage for such crops under ordinary farming conditions. It is necessary after each irrigation to break the soil crust by cultivation to prevent evaporation. This is just as important after irrigation as it is after a rain shower. Also any little pockets that hold water must be carefully drained out, otherwise the crop will be injured by standing water. We are not supposed to have such pockets on land that has been prepared for irrigation.

Kind of Crops to Irrigate.—Wheat, oats, barley, etc., may be helped with one irrigation from imminent failure to a wealth of production. But these rainfall grain crops do not come under the general classification that interests the regular irrigation farmer beyond his diversity plans for producing considerable variety. Fruits, roots, clover, alfalfa, vegetables and Indian corn are money crops under irrigation. Certain seed crops yield splendidly when watered. An apple orchard properly cared for and irrigated just at the right time will pay from five hundred to a thousand dollars per acre. Small fruits are just as valuable. These successes account for the high prices of irrigated land. In the East and in the great Middle West, valuable crops are cut short or ruined by drouth when the fruit or corn is forming. It makes no difference how much rain comes along at other times in the year, if the roots cannot find moisture at the critical time, the yield is reduced often below the profit of raising and harvesting the crop. Strawberry blossoms shrivel and die in the blooming when rain fails. Irrigation is better than rain for strawberries. Strawberries under irrigation may be made to yield more bushels than potatoes under humid conditions. One hundred bushels of strawberries per acre sounds like a fairy tale, but it is possible on rich land under irrigation.

The cost of pumping for irrigation, where the well and machinery is used for no other purpose, must be charged up to the crop. The items of expense are interest on the first cost of the pumping machinery, depreciation, upkeep and running expenses. On Eastern farms, however, where diversified farming is the business, this expense may be divided among the different lines of work. Where live-stock is kept, it is necessary to have a good, reliable water supply for the animals. A reservoir on high ground so water may be piped to the watering troughs and to the house is a great convenience. Also the same engine that does the pumping may be used for other work in connection with the farm, so that the irrigation pump engine, instead of lying idle ten or eleven months in the year, may be utilized to advantage and made to earn its keep. Well-water contains many impurities. For this reason, it is likely to be valuable for crop growing purposes in a wider sense than merely to supply moisture. Well-water contains lime, and lime is beneficial to most soils. It has been noticed that crops grow especially well when irrigated from wells.

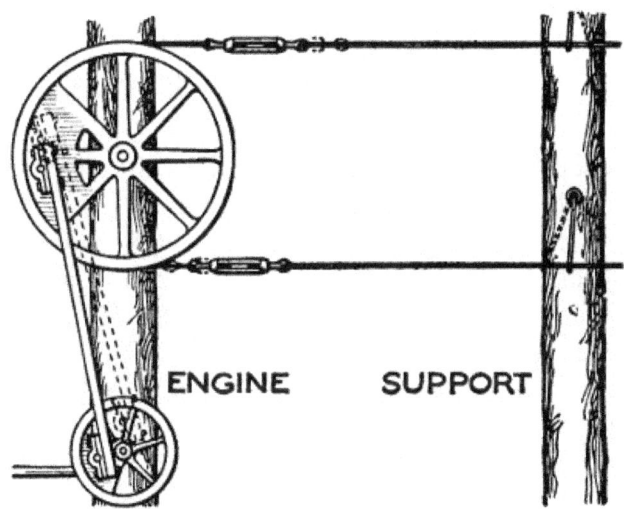

Figure 128.—Power Transmission. Circular motion is converted into reciprocating motion by the different lengths of the two pitman cranks which cause the upper wheel to oscillate. Power is carried to a distance by wires. To reduce friction the wires are supported by swinging hangers. Sometimes wooden rods are used instead of wires to lessen expansion and contraction.

House and Barns Supplied from a Reservoir.—A farm reservoir may sometimes be built very cheaply by throwing a dam across a narrow hollow between two hills, or ridges. On other farms, it is necessary to scrape out a hole on the highest ground within reach. For easy irrigation a reservoir is necessary, and it is economical because the pump may work overtime and supply enough water so the irrigation may be done quickly and with sufficient water to make it effective. When the cost of the reservoir can be charged up to the different departments of the business, such as irrigation, live-stock and house use, the cost is divided and the profits are multiplied.

Power Conveyor.—Circular motion is converted into reciprocal motion to operate a pump at a distance from the engine. The short jack crank oscillates the driving pulley to move the conveyor wires back and forth. The distance to which power may be carried is limited by the expansion and contraction of the conveying wires. Wooden rods are better under extremes of temperature. Where an engine is used night and morning in the dairy house to run a cream separator, this kind of power transmission may be worked to operate the pump at the house. Light wire hangers will support the line wires or rods. They should be about three feet in length, made fast at top and bottom to prevent wear. The spring of a No. 10 wire three feet long is sufficient to swing the length of a pump stroke and the friction is practically nothing.

ELECTRICITY ON THE FARM

Electric current in some sections may be purchased from electric railways or city lighting plants. But the great majority of farms are beyond the reach of high tension transmission cables. In some places three or four farmers may club together and buy a small lighting plant to supply their own premises with both light and power. Unless an engineer is employed to run it trouble is sure to follow, because one family does all of the work and others share equally in the benefits. The solution is for each farmer to install a small plant of his own. The proposition is not so difficult as it sounds. Two-horsepower plants are manufactured for this very purpose. But there is more to it than buying a dynamo and a few lamp bulbs. A farm electric system should supply power to run all of the light stationary machinery about the farm, and that means storage batteries, and the use of one or more small electric motors. There are several ways to arrange the plant, but to save confusion it is better to study first the storage battery plan and to start with an engine large enough to pump water and run the dynamo at the same time. It is a good way to do two jobs at once—you store water enough in the supply tank to last twenty-four or forty-eight hours, and at the same time you store up sufficient electricity to run the cream-separator for a week. Electric power is the only power that is steady enough to get all of the cream.

Figure 129.—Electric Power Plant. A practical farm generator and storage battery, making a complete farm electric plant that will develop and store electricity for instant use in any or all of the farm buildings.

Refrigeration is a profitable way to use electric power. There are small automatic refrigerator machines that maintain low temperatures to preserve food products. This branch of the work may be made profitable. Laundry work on the farm was principally hand labor until the small power washers and wringers were invented. Now a small electric motor takes the blue out of Monday, and the women wear smiles. Electric flatirons afford the greatest comfort on Tuesday. The proper heat is maintained continually until the last piece is ironed. Cooking by electricity is another great success. Some women buy separate cooking utensils, such as toasters, chafing dishes and coffee percolators. Others invest in a regular electric cooking range at a cost of fifty dollars and feel that the money was well spent. It takes about 100 K.W.H. per month in hot weather to cook by electricity for a family of four. In winter, when heat is more of a luxury, the coal or wood range will save half of the electric current. Dishwashing by electricity is another labor-saver three times a day. Vacuum cleaners run by electricity take the dust and microbes out of floor rugs with less hand labor than pushing a carpet sweeper. Incubators are better heated by electricity than any other way. Brooders come under the same class. Sewing-machines were operated by electricity in sweatshops years ago—because it paid. Farm women are now enjoying the same privilege.

Electric lighting on the farm is the most spectacular, if not the most interesting result of electric generation in the country. This feature of the subject was somewhat overtaxed by talkative salesmen representing some of the pioneer manufacturers of electric lighting plants, but the business has steadied down. Real electric generating machinery is being manufactured and sold on its merits in small units.

Not many miles from Chicago there is an electric lighting plant on a dairy farm that is giving satisfaction. The stables are large and they are managed on the plan of milking early in the morning and again in the middle of the afternoon. The morning work requires a great deal of light in the different stables, more light than ordinary, because the milking is done by machinery. The milking machine air-pump is driven by electricity generated on the farm, the power being supplied by a kerosene engine.

Electricity on this farm is used in units, separate lines extending to the different buildings. The lighting plant is operated on what is known as the 32-volt system; the rating costs less to install than some others and the maintenance is less than when a higher voltage is used. I noticed also that there are fewer parts in connection with the plant than in other electric light works that I have examined.

Technical knowledge of electricity and its behavior under different circumstances is hardly necessary to a farmer, because the manufacturers have simplified the mechanics of electric power and lighting to such an extent that it is only necessary to use ordinary precaution to run the plant to its capacity.

At the same time it is just as well to know something about generators, switchboards and the meanings of such terms and names as volt, ampere, battery poles, voltmeter, ammeter, rheostat, discharge switch, underload circuit breaker, false fuse blocks, etc., because familiarity with these names, and the parts they represent gives the person confidence in charging the batteries. Such knowledge also supplies a reason for the one principal battery precaution, which is not to use out all of the electricity the batteries contain.

Those who have electric lighting plants on the farm do not seem to feel the cost of running the plants, because they use the engine for other purposes. Generally manufacturers figure about 1 H.P. extra to run a dynamo to supply from 25 to 50 lights. My experience with farm engines is that for ordinary farm work such as driving the cream separator, working the pump and grinding feed, a two-horse power engine is more useful than any other size. Farmers who conduct business in the usual way will need a three-horsepower engine if they contemplate adding an electric lighting system to the farm equipment.

Among the advantages of an electric lighting system is the freedom from care on the part of the women. There are no lamps to clean or broken chimneys to cut a finger, so that when the system is properly installed the only work the women have to do is to turn the switches to throw the lights on or off as needed.

The expense in starting a farm electric light plant may be a little more than some other installations, but it seems to be more economical in service when figured from a farmer's standpoint, taking into consideration the fact that he is using power for generating electricity that under ordinary farm management goes to waste.

A three-horsepower engine will do the same amount of work with the same amount of gasoline that a two-horsepower engine will do. This statement may not hold good when figured in fractions, but it will in farm practice. Also when running a pump or cream separator the engine is capable of doing a little extra work so that the storage batteries may be charged with very little extra expense.

On one dairy farm a five-horsepower kerosene engine is used to furnish power for various farm purposes. The engine is belted to a direct-current generator of the shunt-wound type. The generator is wired to an electric storage battery of 88 ampere hour capacity. The battery is composed of a number of separate cells. The cells are grouped together in jars. These jars contain the working parts of the batteries. As each jar of the battery is complete in itself, any one jar may be cut out or another added without affecting the other units. The switchboard receives current either from the battery or from the engine and generator direct. There are a number of switches attached to the switchboard, which may be manipulated to turn the current in any direction desired.

Some provision should be made for the renewal of electric lamps. Old lamps give less light than new ones, and the manufacturers should meet customers on some kind of a fair exchange basis. Tungsten lamps are giving good satisfaction for farm use. These lamps are economical of current, which means a reduction of power to supply the same amount of light. The Mazda lamp is another valuable addition to the list of electric lamps.

The Wisconsin *Agriculturist* publishes a list of 104 different uses for electricity on farms. Many of the electrical machines are used for special detail work in dairies where cheese or butter is made in quantity. Sugar plantations also require small units of power that would not apply to ordinary farming. Some of the work mentioned is extra heavy, such as threshing and cutting ensilage. Other jobs sound trivial, but they are all possible labor-savers. Here is the list:

"Oat crushers, alfalfa mills, horse groomers, horse clippers, hay cutters, clover cutters, corn shellers, ensilage cutters, corn crackers, branding irons, currying machines, feed grinders, flailing machines, live stock food warmers, sheep shears, threshers, grain graders, root cutters, bone grinders, hay hoists, clover hullers, rice threshers, pea and bean hullers, gas-electric harvesters, hay balers, portable motors for running threshers, fanning-mills, grain elevators, huskers and shredders, grain drying machines, binder motors, wheat and corn grinders, milking machines, sterilizing milk, refrigeration, churns, cream-separators, butter workers, butter cutting-printing, milk cooling and circulating pumps, milk clarifiers, cream ripeners, milk mixers, butter tampers, milk shakers, curd grinders, pasteurizers, bottle cleaners, bottle fillers, concrete mixers, cider mills, cider presses, spraying machines, wood splitters, auto trucks, incubators, hovers, telephones, electric bells, ice cutters, fire alarms, electric vehicles, electro cultures, water supply, pumping, water sterilizers, fruit presses, blasting magnetos, lighting, interior telephones, vulcanizers, pocket flash lights, ice breakers, grindstones, emery wheels, wood saws, drop hammers, soldering irons, glue pots, cord wood saws, egg testers, burglar alarms, bell ringing transformers, devices for killing insects and pests, machine tools, molasses heaters, vacuum cleaners, portable lamps to attract insects, toasters, hot plates, grills, percolators, flatirons, ranges, toilette articles, water heaters, fans, egg boilers, heating pads, dishwashers, washing machines, curling irons, forge blowers."

GASOLINE HOUSE LIGHTING

Gasoline gas for house lighting is manufactured in a small generator by evaporating gasoline into gas and mixing it with air, about 5 per cent gas and 95 per cent air. We are all familiar with the little brass gasoline torch heater that tinners and plumbers use to heat their soldering irons. The principle is the same.

There are three systems of using gasoline gas for farmhouse lighting purposes, the hollow wire, tube system, and single lamp system.

The hollow wire system carries the liquid gasoline through the circuit in a small pipe called a hollow wire. Each lamp on the circuit takes a few drops of gasoline as needed, converts it into gas, mixes the gas with the proper amount of air and produces a fine brilliant light. Each lamp has its own little generator and is independent of all other lamps on the line.

The tube system of gasoline gas lighting is similar in appearance, but the tubes are larger and look more like regular gas pipes. In the tube system the gas is generated and mixed with air before it gets into the distribution tube, so that lamps do not require separate generators.

In the separate lamp system each lamp is separate and independent. Each lamp has a small supply of gasoline in the base of the lamp and has a gas generator attached to the burner, which converts the gasoline into gas, mixes it with the proper amount of air and feeds it into the burner as required. Farm lanterns are manufactured that work on this principle. They produce a brilliant light.

By investigating the different systems of gasoline gas lighting in use in village stores and country homes any farmer can select the system that fits into his home conditions to the best advantage. In one farmhouse the owner wanted gasoline gas street lamps on top of his big concrete gateposts, and this was one reason why he decided to adopt gasoline gas lighting and to use the separate lamp system.

ACETYLENE GAS

Acetylene lighting plants are intended for country use beyond the reach of city gas mains or electric cables. Carbide comes in lump form in steel drums. It is converted into gas by a generator that is fitted with clock work to drop one or more lumps into water as gas is needed to keep up the pressure. Acetylene gas is said to be the purest of all illuminating gases. Experiments in growing delicate plants in greenhouses lighted with acetylene seem to prove this claim to be correct.

The light also is bright, clear and powerful. The gas is explosive when mixed with air and confined, so that precautions are necessary in regard to using lanterns or matches near the generators. The expense of installing an acetylene plant in a farm home has prevented its general use.

WOOD-SAW FRAMES

There are a number of makes of saw frames for use on farms, some of which are very simple, while others are quite elaborate. Provision usually is made for dropping the end of the stick as it is cut. Sometimes carriers are provided to elevate the blocks onto a pile. Extension frames to hold both ends of the stick give more or less trouble, because when the stick to be sawed is crooked, it is almost impossible to prevent binding. If a saw binds in the kerf, very often the uniform set is pinched out of alignment, and there is some danger of buckling the saw, so that for ordinary wood sawing it is better to have the end of the stick project beyond the jig. If the saw is sharp and has the right set and the right motion, it will cut the stick off quickly and run free while the end is dropping to the ground.

The quickest saw frames oscillate, being supported on legs that are hinged to the bottom of the frame. Oscillating frames work easier than sliding frames. Sliding frames are sometimes provided with rollers, but roller frames are not steady enough. For cross sawing lumber V-shaped grooves are best. No matter what the feeding device is, it should always be protected by a hood over the saw. The frame should fall back of its own weight, bringing the hood with it, so that the saw is always covered except when actually engaged with the stick. Saw-mandrels vary in diameter and length, but in construction they are much alike. For wood sawing the shaft should be $1\frac{3}{8}''$ or $1\frac{1}{2}''$ in diameter. The shaft runs in two babbitted boxes firmly bolted to the saw frame. The frame itself should be well made and well braced.

ROOT PULPER

There are root pulpers with concave knives which slice roots in such a way as to bend the slices and break them into thousands of leafy shreds. The principle is similar to bending a number of sheets of paper so that each sheet will slide past the next one. Animals do not chew roots when fed in large solid pieces. Cattle choke trying to swallow them whole, but they will munch shredded roots with apparent patience and evident satisfaction. American farmers are shy on roots. They do not raise roots in quantities because it requires a good deal of hand labor, but roots make a juicy laxative and they are valuable as an appetizer and they carry mineral. Pulped roots are safe to feed and they offer the best mixing medium for crushed grains and other concentrated foods.

FEED CRUSHER

Instead of grinding grain for feeding, we have what is known as a crusher which operates on the roller-mill principle. It breaks the grains into flour by crushing instead of grinding. It has the advantage of doing good work quickly. Our feed grinding is done in the two-story corncrib and granary. It is one of the odd jobs on the farm that every man likes. The grain is fed automatically into the machine by means of the grain spouts which lead the different kinds of grain down from the overhead bins. The elevator buckets carry the crushed feed back to one of the bins or into the bagger. In either case it is not necessary to do any lifting for the sacks are carried away on a bag truck. We have no use for a scoop shovel except as a sort of big dustpan to use with the barn broom.

STUMP PULLER

Pulling stumps by machinery is a quick operation compared with the old time methods of grubbing, chopping, prying and burning that our forefathers had on their hands. Modern stump pulling machines are small affairs compared with the heavy, clumsy things that were used a few years ago. Some of the new stump pullers are guaranteed to clear an acre a day of ordinary stumpage. This, of course, must be a rough estimate, because stumps, like other things, vary in numbers, size and condition of soundness. Some old stumps may be removed easily while others hang to the ground with wonderful tenacity.

There are two profits to follow the removal of stumps from a partially cleared field. The work already put on the land has in every case cost considerable labor to get the trees and brush out of the way. The land is partially unproductive so long as stumps remain. For this reason, it is impossible to figure on the first cost until the stumps are removed to complete the work and to put the land in condition to raise machine made crops. When the stumps are removed, the value of the land either for selling or for farming purposes is increased at once. Whether sold or farmed, the increasing value is maintained by cropping the land and securing additional revenue.

There are different ways of removing stumps, some of which are easy while others are difficult and expensive. One of the easiest ways is to bore a two-inch auger hole diagonally down into the stump; then fill the auger hole with coal-oil and let it remain for some weeks to soak into the wood. Large stumps may be bored in different directions so the coal-oil will find its way not only through the main part of the stumps but down into the roots. This treatment requires that the stumps should be somewhat dry. A stump that is full of sap has no room for coal-oil, but after the sap partially dries out, then coal oil will fill the pores of the wood. After the stump is thoroughly saturated with coal-oil, it will burn down to the ground, so that the different large roots will be separated. Sometimes the roots will burn below plow depth, but a good heavy pair of horses with a grappling hook will remove the separated roots.

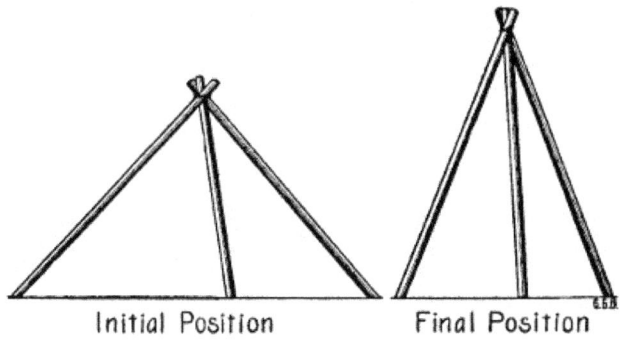

Initial Position   Final Position

Figure 130.—The Oldest Farm Hoist. The first invention for elevating a heavy object was a tripod made of three poles tied together at the top with thongs of bark or rawhide. When hunters were lucky enough to kill a bear, the tripod elevator was erected over the carcass with the lower ends of the poles spread well apart to lower the apex. The gambrel was inserted under the hamstrings and attached to the top of the tripod. As the skinning of the animal proceeded the feet of the tripod were moved closer together. By the time the head was cut off the carcass would swing clear.

Dynamite often is used to blow stumps to pieces, and the work is not considered dangerous since the invention of safety devices. In some sections of the country where firewood is valuable, dynamite has the advantage of saving the wood. An expert with dynamite will blow a stump to pieces so thoroughly that the different parts are easily worked into stove lengths. Pitch-pine stumps have a chemical value that was not suspected until some fellows got rich by operating a retort.

FARM ELEVATING MACHINERY

Many handy and a few heavy elevators are being manufactured to replace human muscle. The simple tripod beef gin was familiar to the early settlers and it is still in use. When a heavy animal was killed for butchering, the small ends of three poles were tied together to form a tripod over the carcass. The feet of the tripod were placed wide apart to raise the apex only a few feet above the animal. After the gambrel was inserted and attached the feet of the tripod were moved gradually closer together as the skinning proceeded, thus elevating the carcass to swing clear of the ground.

Grain Elevators.—As a farm labor-saver, machinery to elevate corn into the two-story corncrib and grain into the upper bins is one of the newer and more important farming inventions. With a modern two-story corncrib having a driveway through the center, a concrete floor and a pit, it is easy to dump a load of grain or ear corn by raising the front end of the wagon box without using a shovel or corn fork. After the load is dumped into the pit a boy can drive a horse around in a circle while the buckets carry the corn or small grain and deliver it by spout into the different corncribs or grain bins. There are several makes of powerful grain elevating machines that will do the work easily and quickly.

The first requisite is a building with storage overhead, and a convenient place to work the machinery. Some of the elevating machines are made portable and some are stationary. Some of the portable machines will work both ways. Usually stationary elevators are placed in vertical position. Some portable elevators may be operated either vertically or on an incline. Such machines are adaptable to different situations, so the corn may be carried up into the top story of a farm grain warehouse or the apparatus may be hauled to the railway station for chuting the grain or ear corn into a car. It depends upon the use to be made of the machinery whether the strictly stationary or portable elevator is required. To unload usually some kind of pit or incline is needed with any kind of an elevator, so the load may be dumped automatically quickly from the wagon box to be distributed by carrying buckets at leisure.

Figure 131.—Portable Grain Elevator Filling a Corncrib. The same rig is taken to the railway to load box cars. The wagon is unloaded by a lifting jack. It costs from 1c to 1½c per bushel to shovel corn by hand, but the greatest saving is in time.

Some elevators are arranged to take grain slowly from under the tailboard of a wagon box. The tailrod is removed and the tailboard raised half an inch or an inch, according to the capacity of the machinery. The load pays out through the opening as the front of the wagon is gradually raised, so the last grain will discharge into the pit or elevator hopper of its own weight. Technical building knowledge and skill is required to properly connect the building and elevating machinery so that the two will work smoothly together. There are certain features about the building that must conform to the requirements and peculiarities of the elevating machinery. The grain and ear corn are both carried up to a point from which they will travel by gravity to any part of the building. The building requires great structural strength in some places, but the material may be very light in others. Hence, the necessity of understanding both building and machinery in order to meet all of the necessary technical requirements.

CHAPTER V

WORKING THE SOIL

IMPORTANCE OF PLOWING

Plowing is a mechanical operation that deals with physics, chemistry, bacteriology and entomology. The soil is the farmer's laboratory; his soil working implements are his mechanical laboratory appliances. A high order of intelligence is required to merge one operation into the next to take full advantage of the assistance offered by nature. The object of plowing and cultivation is to improve the mechanical condition of the soil, to retain moisture, to kill insects and to provide a suitable home for the different kinds of soil bacteria.

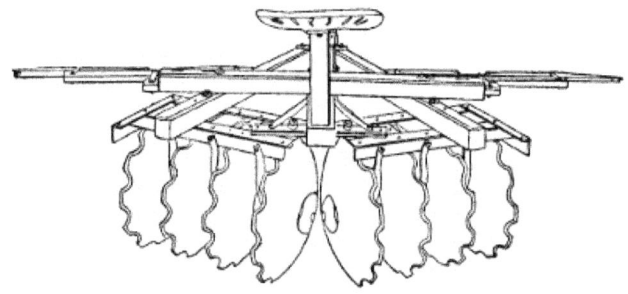

Figure 132.—Heavy Disk Plow. A strong four-horse disk implement for breaking stumpy ground or to tear tough sod into bits before turning under with a moldboard.

There are aerobic and anaerobic bacteria, also nitrogen-gathering bacteria and nitrifying bacteria which are often loosely referred to as azotobacter species. Few of us are on intimate terms with any of them, but some of us have had formal introductions through experiments and observation.

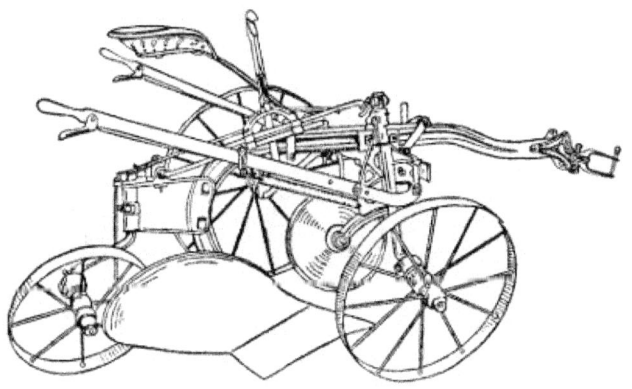

Figure 133.—Sulky Plow. This is a popular type of riding plow. It is fitted with a rolling coulter.

THE MECHANICS OF PLOWING

Walking Plow.—The draft of a walking plow may be increased or diminished by the manner of hitch. It is necessary to find the direct line of draft between the work performed and the propelling force. The clevis in the two-horse doubletree, or the three-horse evener and the adjusting clevis in the end of the plow-beam with the connecting link will permit a limited adjustment. The exact direction that this line takes will prove out in question. The walking plow should not have a tendency to run either in or out, neither too deep nor too shallow. For the proper adjustment as to width and depth of furrow, the plow should follow the line of draft in strict obedience to the pull so that it will keep to the furrow on level ground a distance of several feet without guidance from the handles. In making the adjustment it is first necessary to see that the plow itself is in good working order. All cutting edges such as share, coulter or jointer must be reasonably sharp and the land slip in condition as the makers intended.

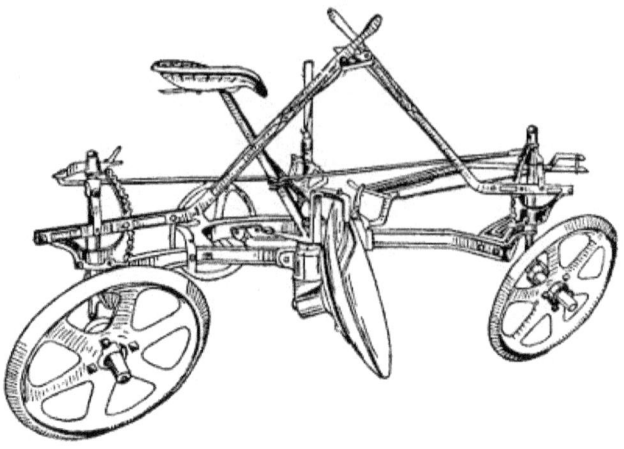

Figure 134.—Disk Plow. Less power is required to plow with a disk, but it is a sort of cut and cover process. The disk digs trenches narrow at the bottom. There are ridges between the little trenches that are not worked.

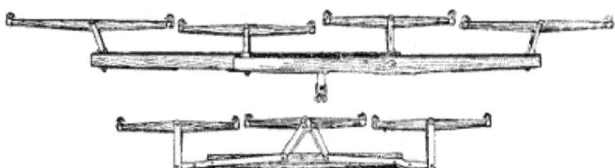

Figure 135.—Three-Horse and Four-Horse Eveners. This kind of evener hitches the horses closer to the load than some others and they are easier to handle than the spread out kinds. The four-horse rig requires the best horses in the middle.

All plows should have a leather pocket on the side of the beam to carry a file. A 12-inch bastard file with a good handle is the most satisfactory implement for sharpening the cutting edges of a plow in the fields. A good deal depends on the character of the soil and its condition of dryness, but generally speaking, it pays to do a little filing after plowing a half mile of furrow. If the horses are doing their duty, a little rest at the end of the half mile is well earned. The plowman can put in the time to advantage with the file and the next half mile will go along merrily in consequence. No farmer would continue to chop wood all day without whetting his axe, but, unfortunately, plowmen often work from morning till night without any attempt to keep the cutting edges of their plows in good working order.

Riding Plow.—The riding plow in lifting and turning the furrow slice depends a good deal on the wheels. The action of the plow is that of a wedge with the power pushing the point, the share and the moldboard between the furrow slices and the land side and the furrow bottom. There is the same friction between the moldboard and the furrow slice as in the case of the walking plow, but the wheels are intended to materially reduce the pressure on the furrow bottom and against the land side. Plow wheels are intended to relieve the draft in this respect because wheels roll much easier than the plow bottom can slide with the weight of the work on top. The track made in the bottom of the furrow with the walking plow shows plainly the heavy pressure of the furrow slice on the moldboard by the mark of the slip. To appreciate the weight the slip carries, an interesting experiment may be performed by loading the walking plow with weights sufficient to make the same kind of a mark when the plow is not turning a furrow.

One advantage in riding plows in addition to the relief of such a load is less packing of the furrow bottom. On certain soils when the moisture is just sufficient to make the subsoil sticky, a certain portion of the furrow bottom is cemented by plow pressure so that it becomes impervious to the passage of moisture either up or down. The track of a plow wheel is less injurious.

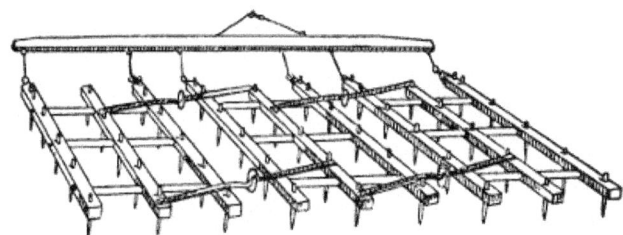

Figure 136.—Three-Section, Spike-Tooth Harrow. The harrow is made straight, but the hitch is placed over to one side to give each tooth a separate line of travel.

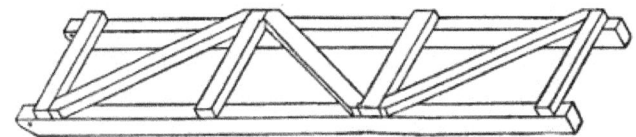

Figure 137.—Harrow Sled Long Enough to Hold a Four-Section Harrow.

Plow wheels should stand at the proper angle to the pressure with especial reference to the work performed. Wheels should be adjusted with an eye single to the conditions existing in the furrow. Some wheel plows apparently are especially built to run light like a wagon above ground regardless of the underground work required of them.

Axles should hang at right angles to the line of lift so accurately as to cause the wheels to wear but lightly on the ends of the hubs. Mistakes in adjustment show in the necessity of keeping a supply of washers on hand to replace the ones that quickly wear thin.

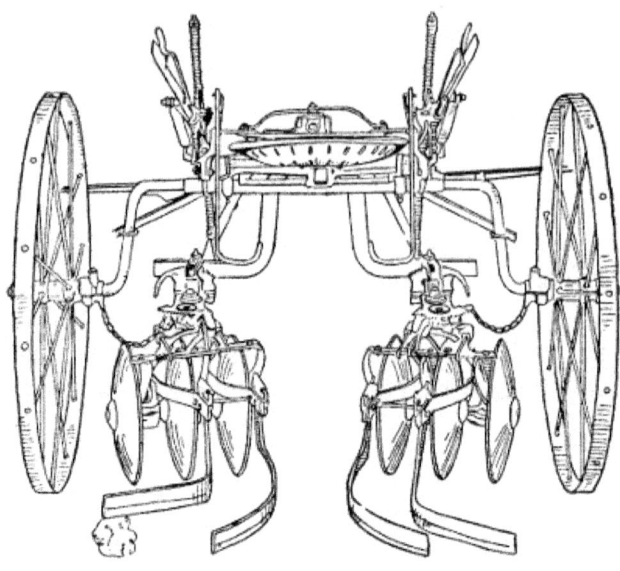

Figure 138.—Corn Cultivator. A one-row, riding-disk cultivator. The ridges are smoothed by the spring scrapers to leave an even surface to prevent evaporation.

In this respect a good deal depends on the sand-bands at the ends of the hubs. Plow wheels are constantly lifting gritty earth and dropping it on the hubs. There is only one successful way to keep sand out of the journals and that is by having the hubs, or hub ferrules, extend well beyond the bearings. Plow wheel hub extensions should reach two inches beyond the journal both at the large end of the hub and at the nut or linchpin end. Some plow wheels cut so badly that farmers consider oil a damage and they are permitted to run dry. This is not only very wasteful of expensive iron but the wheels soon wabble to such an extent that they no longer guide the plow, in which case the draft may be increased enormously.

Figure 139.—A Combination Riding and Walking Cultivator, showing fenders attached to protect young plants the first time through. The two bull tongues shown are for use in heavy soils or when deeper digging is necessary.

Scotch Plows.—When the long, narrow Scotch sod plows are exhibited at American agricultural fairs they attract a good deal of attention and no small amount of ridicule from American farmers because of the six or seven inch furrows they are intended to turn. In this country we are in too much of a hurry to spend all day plowing three-fourths of an acre of ground. Intensive farming is not so much of an object with us as the quantity of land put under cultivation.

Those old-fashioned Scotch plows turn a furrow about two-thirds of the way over, laying the sod surface at an angle of about 45° to the bottom of the furrow. The sharp comb cut by the coulter and share stands upright so that a sod field when plowed is marked in sharp ridges six or seven inches apart, according to the width of the furrow. Edges of sod show in the bottoms of the corrugations between these little furrow ridges.

When the rains come the water is held in these grooves and it finds its way down the whole depth of the furrow slice carrying air with it and moistening every particle of trash clear to the bottom of the furrow. Such conditions are ideal for the work of the different forms of bacteria to break down plant fibre contained in the roots and trash and work it into humus, which is in turn manipulated by other forms of soil bacteria to produce soil water which is the only food of growing plants.

Jointer Plows.—American plow makers also have recognized the necessity of mixing humus with soil in the act of plowing. To facilitate the process and at the same time turn a wide furrow, the jointer does fairly good work when soil conditions are suitable. The jointer is a little plow which takes the place of the coulter and is attached to the plow-beam in the same manner. The jointer turns a little furrow one inch or two inches deep and the large plow following after turns a twelve-inch or fourteen-inch furrow slice flat over, throwing the little jointer furrow in the middle of the furrow bottom in such a way that the big furrow breaks over the smaller furrow.

If the work is well done, cracks as wide as a man's hand and from three to five inches deep are left all over the field. These cracks lead air and moisture to rot the trash below. This is a much quicker way of doing a fairly good job of plowing. Such plows loosen the soil and furnish the conditions required by nature; and they may be operated with much less skill than the old-fashioned narrow-furrowed Scotch plows.

Good plowing requires first that the soil be in proper condition to plow, neither too dry nor too wet, but no man can do good plowing without the proper kind of plow to fit the soil he is working with.

PLOWING BY TRACTOR

Under present conditions farm tractors are not intended to replace horse power entirely but to precede horses to smooth the rough places that horses may follow with the lighter machines to add the finishing touches. Light tractors are being made, and they are growing in popularity, but the real business of the farm tractor is to do the heavy lugging—the work that kills horses and delays seeding until the growing season has passed. The actual power best suited to the individual farm can only be determined by the nature of the land and the kind of farming.

In the Middle West where diversified farming is practiced, the 8-16 and the 10-20 sizes seem to be the most satisfactory, and this is without regard to the size of the farm. The preponderance of heavy work will naturally dictate the buying of a tractor heavier than a 10-20. The amount of stationary work is a factor. In certain communities heavy farm tractors are made to earn dividends by running threshing machines after harvest, silo fillers in the fall and limestone crushers in the winter.

Here is a classified list of jobs the medium size farm tractor is good for:

Clearing the Land—pulling up bushes by the roots, tearing out hedges, pulling stumps, grubbing, pulling stones.

Preparing Seed Bed and Seeding—plowing, disking, crushing clods, pulling a land plane, rolling, packing, drilling, harrowing.

Harvesting—mowing, pulling grain binders, pulling potato digger.

Belt Work—hay baling, corn shelling, heavy pumping for irrigation, grinding feed, threshing, clover hulling, husking and shredding, silo filling, stone crushing.

Road Work—grading, dragging, leveling, ditching, hauling crops.

Miscellaneous—running portable sawmill, stretching wire fencing, ditch digging, manure spreading.

Generally speaking, however, the most important farm tractor work is preparing the seed-bed thoroughly and quickly while the soil and weather conditions are the best. And the tractor's ability to work all day and all night at such times is one of its best qualifications.

To plow one square mile, or 640 acres, with a walking plow turning a twelve-inch furrow, a man and team must walk 5,280 miles. The gang-plow has always been considered a horse killer, and, when farmers discovered that they could use oil power to save their horses, many were quick to make the change.

It requires approximately 10 horsepower hours to turn an acre of land with horses. At a speed of two miles, a team with one plow in ten hours will turn two acres. To deliver the two horsepower required to do this work, they must travel 176 feet per minute and exert a continuous pull of 375 pounds or 187.5 pounds per horse.

One horsepower equals a pull of 33,000 pounds, moved one foot per minute. Two-mile speed equals two times 5,280 or 10,560 feet per hour, or 176 feet per minute. Sixty-six thousand divided by 176 equals 375 foot pounds pull per minute. One horsepower is absorbed in 88 feet of furrow.

Horse labor costs, according to Government figures, 12½ cents per hour per horse. On this basis ten hours' work will be $1.25, which is the average daily cost of each horse. An average Illinois diversified farm of 160 acres would be approximately as follows: Fifty acres of corn, 30 acres of oats and wheat, 20 acres of hay, 60 acres of rough land, pasture, orchard, building and feed lots.

This average farm supports six work horses or mules and one colt. According to figures taken from farm work reports submitted by many different corn belt farmers, the amount of horse-work necessary to do this cropping would figure out as follows:

Fifty acres of corn land for plowing, disking, harrowing, planting, cultivating and harvesting would amount to a total of 1,450 horsepower hours. Thirty acres of wheat would require a total of 330 horsepower hours. Twenty acres of hay would require 110 horsepower hours. In round figures, 1,900 horsepower hours at 12½ cents would amount to $237.50.

Elaborate figures have been worked out theoretically to show that this work can be done by an 8-16 farm tractor in 27¾ days at a cost for kerosene fuel and lubricating oil of $1.89 per day. Adding interest, repairs and depreciation, brings this figure up to about $4.00 per day, or a total of $111.00 for the job. No account is kept of man power in caring for either the horses or the tractor. The actual man labor on the job, however, figures 12⅓ days less for the tractor than for horses. We should remember that actual farm figures are used for the cost of horse work. Such figures are not available for tractor work.

The cost of plowing with a traction engine depends upon so many factors that it is difficult to make any definite statement. It depends upon the condition of the ground, size of the tractor, the number of plows pulled, and the amount of fuel used. An 8-16 horsepower tractor, for instance, burning from 15 to 20 gallons of low grade kerosene per ten hour day and using one gallon of lubricating oil, costs about $1.90 per ten hours work. Pulling two 14-inch plows and traveling 20 miles per day, the tractor will plow 5.6 acres at a fuel and an oil cost of about 30 cents per acre. Pulling three 14-inch plows, it will turn 8.4 acres at a cost for fuel and oil of about 20 cents an acre.

The kind and condition of soil is an important factor in determining the tractor cost of plowing. Comparison between the average horse cost and the average tractor cost suggests very interesting possibilities in favor of tractor plowing under good management.

Aside from the actual cost in dollars we should also remember that no horse gang can possibly do the quality of work that can be accomplished by an engine gang. Anxiety to spare the team has cut a big slice off the profits of many a farmer. He has often plowed late on account of hard ground, and he has many times allowed a field to remain unplowed on account of worn-out teams. Under normal conditions, late plowing never produces as good results as early plowing. Many a farmer has fed and harnessed by the light of the lantern, gone to the field and worked his team hard to take advantage of the cool of the morning. With the approach of the hot hours of midday, the vicious flies sapping the vitality from his faithful team, he has eased up on the work or quit the job.

In using the tractor for plowing, there are none of these distressing conditions to be taken into consideration, nothing to think of but the quality of work done. It is possible to plow deep without thought of the added burden. Deep plowing may or may not be advisable. But where the soil will stand it, deep plowing at the proper time of year, and when done with judgment, holds moisture better and provides more plant food.

The pull power required to plow different soils varies from about three pounds per square inch of furrow for light sand up to twenty pounds per square inch of furrow for gumbo. The draft of a plow is generally figured from clover sod, which averages about seven pounds per square inch. Suppose a plow rig has two 14-inch bottoms, and the depth to be plowed is six inches. A cross section of each plow is therefore 14 by 6 inches, or 84 square inches. Twice this for two bottoms is 168 square inches. Since, in sandy soil, the pressure per square inch is three pounds, therefore 168 times 3 pounds equals 504 pounds, the draft in sandy soil. 168 times 7 pounds equals 1,176 pounds, the draft in clover sod. 168 times 8 pounds equals 1,344 pounds, the draft in clay sod.

The success of crop growing depends upon the way the seed-bed is prepared. The final preparation of the seed-bed can never be thoroughly well done unless the ground is properly plowed to begin with. It is not sufficient to root the ground over or to crowd it to one side but the plow must really turn the furrow slice in a uniform, systematic manner and lay it bottom side uppermost to receive the beneficial action of the air, rain and sunshine.

The moldboard of a plow must be smooth in order to properly shed the earth freely to make an easy turn-over. The shape of the shear and the forward part of the moldboard is primarily that of a wedge, but the roll or upper curve of the moldboard changes according to soil texture and the width and depth of furrow to be turned. Moldboards also differ in size and shape, according to the kind of furrow to be turned. Sometimes in certain soils a narrow solid furrow with a comb on the upper edge is preferable. In other soils a cracked or broken furrow slice works the best. When working our lighter soils a wide furrow turned flat over on top of a jointer furrow breaks the ground into fragments with wide cracks or openings reaching several inches down. Between these extremes there are many modifications made for the particular type or texture of the soil to be plowed. We can observe the effect that a rough, or badly scratched, or poorly shaped moldboard has on any kind of soil, especially when passing from gravelly soils to clay. In soil that contains the right amount of moisture, when a plow scours all the time, the top of the furrow slice always has a glazed or shiny appearance. This shows that the soil is slipping off the moldboard easily. In places where the plow does not scour the ground is pushed to one side and packed or puddled on the underside instead of being lifted and turned as it should be. A field plowed with a defective moldboard will be full of these places. Such ground cannot have the life to bring about a satisfactory bacteria condition necessary to promote the rapid plant growth that proper plowing gives it.

Cultivated sandy soils are becoming more acid year after year. We are using lime to correct the acidity, but the use of lime requires better plowing and better after cultivation to thoroughly mix the trash with the earth to make soil conditions favorable to the different kinds of soil bacteria. Unless we pay special attention to the humus content of the soil we are likely to use lime to dissolve out plant foods that are not needed by the present crop, and, therefore, cannot be utilized. This is what the old adage means which reads: "Lime enricheth the father but impoverisheth the son." When that was written the world had no proper tillage tools and the importance of humus was not even dreamed of.

Not so many years ago farm plows were made of cast iron. Then came the steel moldboard, which was supposed to be the acme of perfection in plow making. Steel would scour and turn the furrow in fluffy soils where cast iron would just root along without turning the ground at all. Later the art of molding steel was studied and perfected until many grades and degrees of hardness were produced and the shape of the moldboard passed through a thousand changes. The idea all the time was to make plows that would not only scour but polish in all kinds of soil. At the same time they must turn under all of the vegetable growth to make humus, to kill weeds and to destroy troublesome insects. Besides these requirements the soil must be pulverized and laid loose to admit both air and moisture. These experiments gradually led up to our present high grade plows of hardened steel and what is known as chilled steel.

Besides the hardness there are different shapes designed for different soils so that a plow to work well on one farm may need to be quite different from a plow to do the best work in another neighborhood. The furrow slice sliding over a perfect moldboard leaves the surface of the upturned ground as even as the bottom of the furrow. By using a modern plow carefully selected to fit the soil, gravel, sandy, stony or muck soils, or silt loams that contain silica, lime, iron and aluminum oxide can be worked with the right plow to do the best work possible if we use the necessary care and judgment in making the selection.

One object of good plowing is to retain moisture in the soil until the growing crop can make good use of it.

The ease with which soils absorb, retain or lose moisture, depends mostly on their texture, humus content, physical condition, and surface slope or artificial drainage. It is to the extent that cultivation can modify these factors that more soil water can be made available to the growing crop. There are loose, open soils through which water percolates as through a sieve, and there are tight, gumbo soils which swell when the surface is moistened and become practically waterproof. Sandy soils take in water more readily than heavier soils, hence less precaution is necessary to prevent run-off.

Among the thousands of plows of many different makes there are plenty of good ones. The first consideration in making a selection is a reliable home dealer who has a good business reputation and a thorough knowledge of local soil from a mechanical standpoint. The next consideration is the service the plow will give in proportion to the price.

DISK HARROW

For preparing land to receive the seed no other implement will equal a double disk. These implements are made in various sizes and weights of frame. For heavy land, where it is necessary to weight the disk down, an extra heavy frame is necessary. It would probably be advisable to get the extra strong frame for any kind of land, because even in light sand there are times when a disk may be used to advantage to kill quackgrass or to chew up sod before plowing. In such cases it is customary to load on a couple of sacks of sand in addition to the weight of the driver. When a disk is carrying 300 or 400 pounds besides its own weight the racking strains which pull from different directions have a tendency to warp or twist a light frame out of shape. To keep a disk cultivator in good working order it is necessary to go over it thoroughly before doing heavy work. Bolts must be kept tight, all braces examined occasionally, and the heavy nuts at the ends of the disk shafts watched. They sometimes loosen and give trouble. The greatest difficulty in running a disk harrow or cultivator is to keep the boxings in good trim. Wooden boxes are provided with the implement. It is a good plan to insist on having a full set of eight extra boxes. These wooden boxes may be made on the farm, but it sometimes is difficult to get the right kind of wood. They should be made of hard maple, bored according to size of shaft, and boiled in a good quality of linseed oil. Iron boxings have never been satisfactory on a disk implement. Wooden ones make enough trouble, but wood has proved better than iron.

On most disk cultivators there are oil channels leading to the boxings. These channels are large enough to carry heavy oil. The lighter grades of cylinder oil work the best. It is difficult to cork these oil channels tight enough to keep the sand out. Oil and sand do not work well together in a bearing. The manufacturers of these implements could improve the oiling device by shortening the channel and building a better housing for the oil entrance. It is quite a job to take a disk apart to put in new boxings, but, like all other repair work, the disk should be taken into the shop, thoroughly cleaned, repaired, painted and oiled in the winter time.

Some double disk cultivators have tongues and some are made without. Whether the farmer wants a tongue or not depends a good deal on the land. The only advantage is that a tongue will hold the disk from crowding onto the horses when it is running light along the farm lanes or the sides of the fields with the disks set straight. Horses have been ruined by having the sharp disks run against them when going down hill. Such accidents always are avoidable if a man realizes the danger. Unfortunately, farm implements are often used by men who do very little thinking. A spring disk scraper got twisted on a root and was thrown over the top of one of the disks so it scraped against the back of the disk and continued to make a harsh, scraping noise until the proprietor went to see what was wrong. The man driving the disk said he thought something must be the matter with the cultivator, but he couldn't tell for the life of him what it was. When farmers are up against such difficulties it is safer to buy a disk with a tongue.

Harrow Cart.—A small two-wheel cart with a spring seat overshadowed with a big umbrella is sometimes called a "dude sulky." Many sensitive farmers trudge along in the soft ground and dust behind their harrows afraid of such old fogy ridicule. The hardest and most tiresome and disagreeable job at seeding time is following a harrow on foot. Riding a harrow cart in the field is conserving energy that may be applied to better purposes after the day's work in the field is finished.

KNIFE-EDGE PULVERIZERS

A knife-edge weeder makes the best dust mulch pulverizer for orchard work or when preparing a seed-bed for grain. These implements are sold under different names. It requires a stretch of imagination to attach the word "harrow" to these knife-edge weeders. There is a central bar which is usually a hardwood plank. The knives are bolted to the underside of the plank and sloped backward and outward from the center to the right and left, so that the knife-edges stand at an angle of about 45° to the line of draught. This angle is just about sufficient to let tough weeds slip off the edges instead of dragging along. If the knives are sharp, they will cut tender weeds, but the tough ones must be disposed of to prevent choking. The proper use of the knife-edge weeder prevents weeds from growing, but in farm practice, sometimes rainy weather prevents the use of such a tool until the weeds are well established. As a moisture retainer, these knife-edge weeders are superior to almost any other implement. They are made in widths of from eight to twenty feet. The wide ones are jointed in the middle to fit uneven ground.

CLOD CRUSHER

The farm land drag, float, or clod crusher is useful under certain conditions on low spots that do not drain properly. Such land must be plowed when the main portion of the field is in proper condition, and the result often is that the low spots are so wet that the ground packs into lumps that an ordinary harrow will not break to pieces. Such lumps roll out between the harrow teeth and remain on top of the ground to interfere with cultivation. The clod crusher then rides over the lumps and grinds them into powder. Unfortunately, clod crushers often are depended on to remedy faulty work on ordinary land that should receive better treatment. Many times the clod crusher is a poor remedy for poor tillage on naturally good land that lacks humus.

Figure 140.—Land Float. Clod crushers and land floats belong to the same tribe. Theoretically they are all outlaws, but some practical farmers harbor one or more of them. Wet land, containing considerable clay, sometimes forms into lumps which should be crushed.

As ordinarily made, the land float or clod crusher consists of from five to eight planks, two inches thick and ten or twelve inches wide, spiked together in sawtooth position, the edges of the planks being lapped over each other like clapboards in house siding. The planks are held in place with spikes driven through into the crosspieces.

FARM ROLLER

Farm rollers are used to firm the soil. Sometimes a seed-bed is worked up so thoroughly that the ground is made too loose so the soil is too open and porous. Seeds to germinate require that the soil grains shall fit up closely against them. Good soil is impregnated with soil moisture, or film moisture as it is often called, because the moisture forms in a film around each little soil grain. In properly prepared soil this film moisture comes in contact with the freshly sown seed. If the temperature is right the seed swells and germination starts. The swelling of the seed brings it in contact with more film moisture attached to other grains of soil so the rootlet grows and pushes out into the soil in search of moisture on its own account. A roller is valuable to press the particles of soil together to bring the freshly sown seeds in direct contact with as many particles of soil as possible. Rolling land is a peculiar operation, the value of which is not always understood. The original idea was to benefit the soil by breaking the lumps. It may be of some benefit on certain soils for this purpose, but the land should always be harrowed after rolling to form a dust mulch to prevent the evaporation of moisture. Land that has been rolled and left overnight shows damp the next morning, which is sufficient proof that moisture is coming to the surface and is being dissipated into the atmosphere. In the so-called humid sections of the country the great problem is to retain moisture. Any farm implement that has a tendency to dissipate soil moisture is a damage to the farmer. Probably nine times out of ten a farm roller is a damage to the crop it is intended to benefit because of the manner in which it is used. It is the abuse, not the proper use of a roller, that injures the crop.

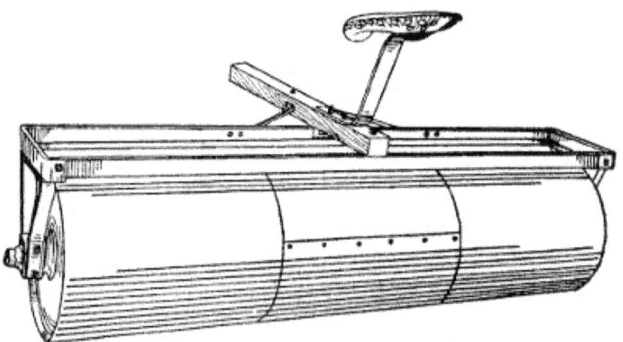

Figure 141.—Iron Land Roller Made of Boiler Plate.

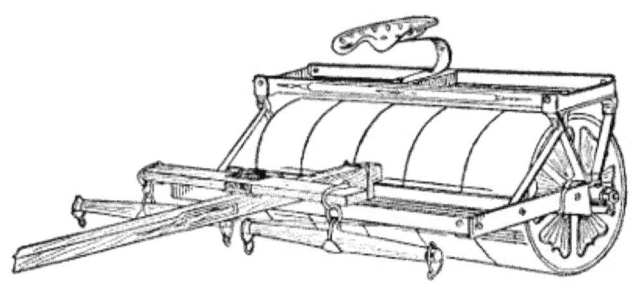

Figure 142.—Wooden Land Roller.

CORN-PLANTER

Corn-planters are designed to plant two rows at once. The width of rows may be adjusted from about 32 to 44 inches apart. When seed-corn is carefully graded to size the dropping mechanism will feed out the grains of corn regularly with very few skips. This is one reason why most farmers plant corn in drills. There are other cultural reasons which do not properly belong to this mechanical article. Hill dropping is considerably more complicated and difficult. After the feeding mechanism has been adjusted to the size of seed kernels to be planted so it will drop four kernels in a hill then the trip chain is tried out to see if it is right at every joint. Dropping in hills is a very careful mechanical proposition. An inch or two out of line either way means a loss of corn in cultivating.

In setting the stakes to go and come by, a careful measurement of the field is necessary in order to get the stake lines on both sides of the field parallel. If the ring stakes are driven accurately on the line, then the first hill of corn must come at the same distance from the line in each row. Likewise in starting back from the far side of the field the first hill should measure exactly the same distance from the stake line as the first hills on the opposite side of the field. This is easily managed by counting the number of trips between the stake line and the first row of corn hills. If the two lines of stakes on the opposite sides of the field are exactly parallel it is not necessary to move either line in order to get the proper distance to start dropping, but it must be adjusted by measurement, otherwise the corn hills will be dodged. If the corn hills are to space three feet apart then the first row of hills should come nine or twelve feet from the stake line. Stakes may be measured and set a certain number of inches from the line to make the distance come right. This careful adjustment brings the hills in line in the rows.

When the field is level or gently sloping there is no difficulty in making straight rows so far as check rowing is concerned. When the field is hilly another problem crops up. It is almost impossible to run corn rows along the side of a hill and keep them straight. The planter has a tendency to slide downhill. Also the distance across a field is greater where the rows pass over a hill. To keep the rows straight under such conditions allowance must be made for the stretch over the hill as well as for the side thrust of the planter. Where a chain marker is used it hangs downhill and a further allowance must be made for that. A good driver will skip an inch or so above the mark so that the rows will be planted fairly straight. This means a good deal more in check rowing than when the corn is planted in drills. The greatest objection to hill planting is the crowding of four corn plants into a space that should be occupied by one plant.

A great many experiments have been tried to scatter the seeds in the hill, so far without definite results, except when considerable additional expense is incurred. However, a cone suspended below the end of the dropping tube usually will scatter the seeds so that no two seeds will touch each other. They may not drop and scatter four or five inches apart, but these little cones will help a good deal. They must be accurately adjusted so the point of the cone will center in the middle of the vertical delivery tube, and there must be plenty of room all around the cone so the corn seed kernels won't stick. The braces that hold the cones in place for the same reason must be turned edge up and supported in such a way as to leave plenty of clearance. The idea is that four kernels of corn drop together. They strike the cone and are scattered in different directions. They naturally fly to the outsides of the drill mark which scatters them as wide apart as the width of the shoe that opens the drill. The advantage of scattering seed grains in the hill has been shown by accurate experiments conducted at different times by agricultural colleges.

GRAIN DRILL

To know exactly how much seed the grain drill is using it is necessary to know how many acres are contained in the field. Most drills have an attachment that is supposed to measure how many acres and fractions of acres the drill covers. Farmers know how much grain each sack contains, so they can estimate as they go along, provided the drill register is correct. It is better to provide a check on the drill indicator. Have the field measured, then drive stakes along one side, indicating one acre, five acres and ten acres. When the one-acre stake is reached the operator can estimate very closely whether the drill is using more or less seed than the indicator registers. When the five-acre stake is reached another proof is available, and so on across the field. Next in importance to the proper working of the drill is straight rows. The only way to avoid gaps is to drive straight. The only way to drive straight is to sight over the wheel that follows the last drill mark. Farmers sometimes like to ride on the grain drill, which places the wheel sighting proposition out of the question. A harrow cart may be hitched behind the wheel of the grain drill, but it gives a side draft. The only way to have straight rows and thorough work is to walk behind the end of the drill. This is the proper way to use a drill, anyway, because a tooth may clog up any minute. Unless the operator is walking behind the drill he is not in position to see quickly whether every tooth is working properly or not. It is hard work to follow a drill all day long, but it pays at harvest time. It costs just as much to raise a crop of grain that only covers part of the ground, and it seems too bad to miss the highest possible percentage to save a little hard work at planting time.

SPECIAL CROP MACHINERY

Special crops require special implements. After they are provided, the equipment must be kept busy in order to make it pay. If a farmer produces five acres of potatoes he needs a potato cutter, a planter, a riding cultivator, a sprayer that works under high pressure, a digger and a sorter. The same outfit will answer for forty acres, which would reduce the per acre cost considerably. No farmer can afford to grow five acres of potatoes without the necessary machinery, because hand labor is out of the question for work of that kind.

On the right kind of soil, and within reach of the right market, potatoes are money-makers. But they must be grown every year because the price of potatoes fluctuates more than any other farm crop. Under the right conditions potatoes grown for five years with proper care and good management are sure to make money. One year out of five will break even, two years will make a little money and the other two years will make big money. At the end of five years, with good business management, the potato machinery will be all paid for, and there will be a substantial profit.

WHEEL HOE

In growing onions and other truck crops, where the rows are too close together for horse cultivation, the wheel hoe is valuable. In fact, it is almost indispensable when such crops are grown extensively. The best wheel hoes have a number of attachments. When the seed-bed has been carefully prepared, and the soil is fine and loose, the wheel hoe may be used as soon as the young plants show above ground. Men who are accustomed to operating a wheel hoe become expert. They can work almost as close to the growing plants with an implement of this kind as they can with an ordinary hand hoe. The wheel hoe, or hand cultivator, works the ground on both sides of the row at once, and it does it quickly, so that very little hand weeding is necessary.

CHAPTER VI

HANDLING THE HAY CROP

REVOLVING HAYRAKE

About the first contrivance for raking hay by horse power consisted of a stick eight or ten feet long with double-end teeth running through it, and pointing in two directions. These rakes were improved from time to time, until they reached perfection for this kind of tool. They have since been superseded by spring-tooth horse rakes, except for certain purposes. For pulling field peas, and some kinds of beans, the old style revolving horse rake is still in use.

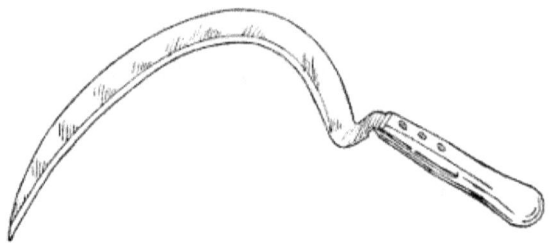

Figure 143.—Grass Hook, for working around borders where the lawn-mower is too clumsy.

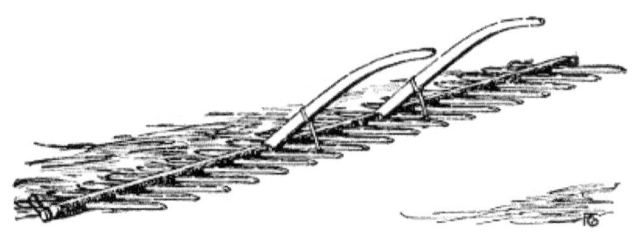

Figure 144.—Revolving Hayrake. The center piece is 4" x 6" x 12' long. The teeth are double enders 1⅜" square and 4' 6" long, which allows 24" of rake tooth clear of the center timber. Every stick in the rake is carefully selected. It is drawn by one horse. If the center teeth stick into the ground either the horse must stop instantly, or the rake must flop over, or there will be a repair job. This invention has never been improved upon for pulling Canada peas.

Improved revolving horse rakes have a center timber of hardwood about 4 x 6 inches in diameter. The corners are rounded to facilitate sliding over the ground. A rake twelve feet long will have about eighteen double-end teeth. The teeth project about two and one-half feet each way from the center timber. Each tooth is rounded up, sled-runner fashion, at each end so it will point forward and slide along over and close to the ground without catching fast. There is an iron pull rod, or long hook, attached to each end of the center bar by means of a bolt that screws into the center of the end of the wooden center shaft, thus forming a gudgeon pin so the shaft can revolve. Two handles are fastened by band iron straps to rounded recesses or girdles cut around the center bar. These girdles are just far enough apart for a man to walk between and to operate the handles. Wooden, or iron lugs, reach down from the handles with pins projecting from their sides to engage the rake teeth. Two pins project from the left lug and three from the right. Sometimes notches are made in the lugs instead of pins. Notches are better; they may be rounded up to prevent catching when the rake revolves. As the rake slides along, the driver holds the rake teeth in the proper position by means of the handles. When sufficient load has been gathered he engages the upper notch in the right hand lug, releases the left and raises the other sufficient to point the teeth into the ground. The pull of the horse turns the rake over and the man grasps the teeth again with the handle lugs as before. Unless the driver is careful the teeth may stick in the ground and turn over before he is ready for it. It requires a little experience to use such a rake to advantage. No better or cheaper way has ever been invented for harvesting Canada peas. The only objections are that it shells some of the riper pods and it gathers up a certain amount of earth with the vines which makes dusty threshing.

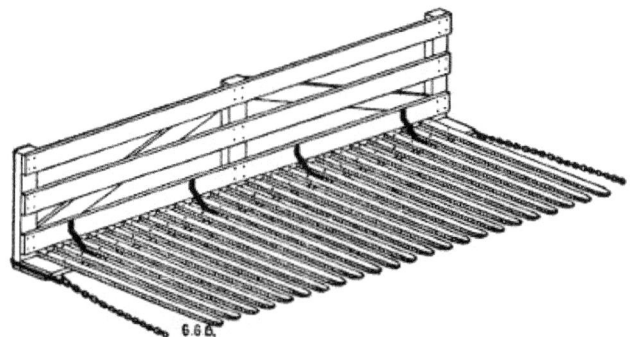

Figure 145.—Buck Rake. When hay is stacked in the field a four-horse buck rake is the quickest way to bring the hay to the stack. The buck rake shown is 16 feet wide and the 2 x 4 teeth are 11 feet long. Two horses are hitched to each end and two drivers stand on the ends of the buck rake to operate it. The load is pushed under the horse fork, the horses are swung outward and the buck rake is dragged backward.

HAY-TEDDER

The hay-tedder is an English invention, which has been adopted by farmers in rainy sections of the United States. It is an energetic kicker that scatters the hay swaths and drops the hay loosely to dry between showers. Hay may be made quickly by starting the tedder an hour behind the mowing machine.

It is quite possible to cut timothy hay in the morning and put it in the mow in the afternoon, by shaking it up thoroughly once or twice with the hay-tedder. When clover is mixed with the timothy, it is necessary to leave it in the field until the next day, but the time between cutting and mowing is shortened materially by the use of the tedder.

Grass cut for hay may be kicked apart in the field early during the wilting process without shattering the leaves. If left too long, then the hay-tedder is a damage because it kicks the leaves loose from the stems and the most valuable feeding material is wasted. But it is a good implement if rightly used. In catchy weather it often means the difference between bright, valuable hay and black, musty stuff, that is hardly fit to feed.

Hay-tedders are expensive. Where two farmers neighbor together the expense may be shared, because the tedder does its work in two or three hours' time. Careful farmers do not cut down much grass at one time. The tedder scatters two mowing swaths at once. In fact the mowing machine, hay-tedder and horserake should all fit together for team work so they will follow each other without skips or unnecessary laps. The dividing board of the mowing-machine marks a path for one of the horses to follow and it is difficult to keep him out of it. But two horses pulling a hay-tedder will straddle the open strip between the swaths when the tedder is twice the width of the cut.

HAY SKIDS

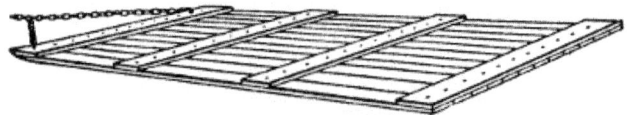

Figure 146.—Hay Skid. This hay skid is 8 feet wide and 16 feet long. It is made of ⅞" lumber put together with 2" carriage bolts—plenty of them. The round boltheads are countersunk into the bottom of the skid and the nuts are drawn down tight on the cleats. It makes a low-down, easy-pitching, hay-hauling device.

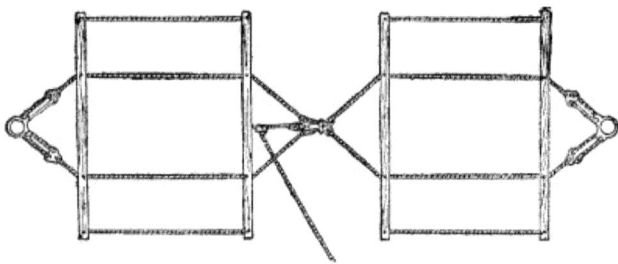

Figure 147.—Hay Sling. It takes no longer to hoist 500 pounds of hay than 100 pounds if the rig is large and strong enough. Four feet wide by ten feet in length is about right for handling hay quickly. But the toggle must reach to the ends of the rack if used on a wagon.

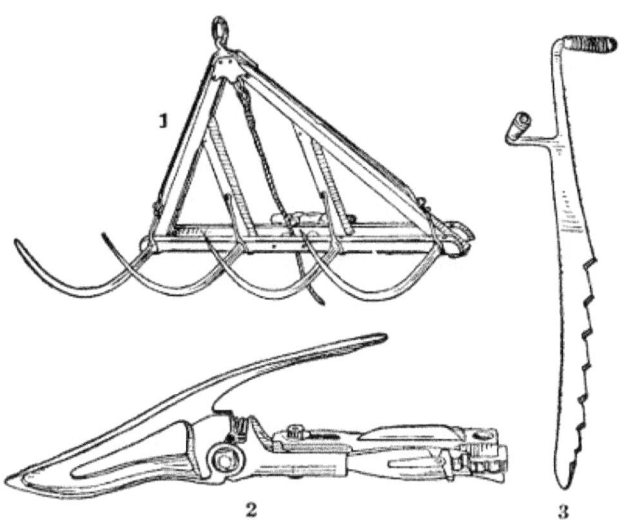

Figure 148.—(1) Four-Tined Derrick Fork. (2) Pea Guard. An extension guard to lift pea-vines high enough for the sickle is the cleanest way to harvest Canada peas. The old-fashioned way of pulling peas with a dull scythe has gone into oblivion. But the heavy bearing varieties still persist in crawling on the ground. If the vines are lifted and cut clean they can be raked into windrows with a spring tooth hayrake. (3) Haystack Knife. This style of hay-cutting knife is used almost universally on stacks and in hay-mows. There is less use for hay-knives since farmers adopted power hayforks to lift hay out of a mow as well as to put it in.

Hay slips, or hay skids, are used on the old smooth fields in the eastern states. They are usually made of seven-eighths-inch boards dressed preferably on one side only. They are used smooth side to the ground to slip along easily. Rough side is up to better hold the hay from slipping. The long runner boards are held together by cross pieces made of inch boards twelve inches wide and well nailed at each intersection with nails well clinched. Small carriage bolts are better than nails but the heads should be countersunk into the bottom with the points up. They should be used without washers and the ends of the bolts cut close to the sunken nuts. The front end of the skid is rounded up slightly, sled runner fashion, as much as the boards will bear, to avoid digging into the sod to destroy either the grass roots or crowns of the plants. Hay usually is forked by hand from the windrows on to the skids. Sometimes hay slings are placed on the skids and the hay is forked on to the slings carefully in layers lapped over each other in such a way as to hoist on to the stack without spilling out at the sides. Four hundred to eight hundred pounds makes a good load for one of these skids, according to horse power and unevenness of the ground. They save labor, as compared to wagons, because there is no pitching up. All hoisting is supposed to be done by horse power with the aid of a hay derrick.

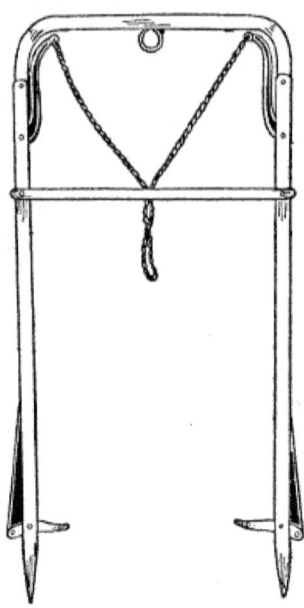

Figure 149.—Double Harpoon Hayfork. This is a large size fork with extra long legs. For handling long hay that hangs together well this fork is a great success. It may be handled as quickly as a smaller fork and it carries a heavy load.

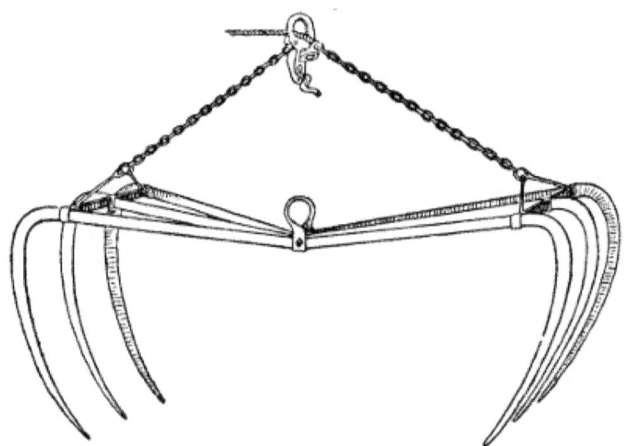

Figure 150.—Six-Tined Grapple Hayfork. It is balanced to hang as shown in the drawing when empty. It sinks into the hay easily and dumps quickly when the clutch is released.

WESTERN HAY DERRICKS

Two derricks for stacking hay, that are used extensively in the alfalfa districts of Idaho, are shown in the illustration, Figure 151. The derrick to the left is made with a square base of timbers which supports an upright mast and a horizontal boom. The timber base is sixteen feet square, made of five sticks of timber, each piece being 8 x 8 inches square by 16 feet in length. Two of the timbers rest flat on the ground and are rounded up at the ends to facilitate moving the derrick across the stubble ground or along the road to the next hayfield. These sleigh runner timbers are notched on the upper side near each end and at the middle to receive the three cross timbers. The cross timbers also are notched or recessed about a half inch deep to make a sort of double mortise. The timbers are bound together at the intersections by iron U-clamps that pass around both timbers and fasten through a flat iron plate on top of the upper timbers. These flat plates or bars have holes near the ends and the threaded ends of the U-irons pass through these holes and the nuts are screwed down tight. The sleigh runner timbers are recessed diagonally across the bottom to fit the round U-irons which are let into the bottoms of the timbers just enough to prevent scraping the earth when the derrick is being moved. These iron U-clamp fasteners are much stronger and better than bolts through the timbers.

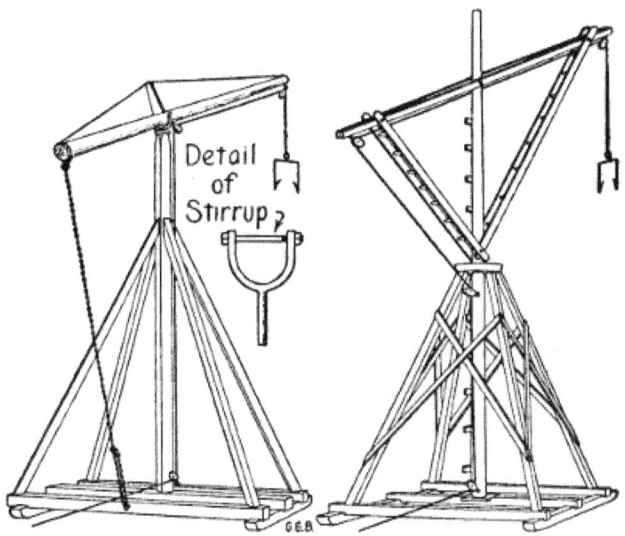

Figure 151.—Idaho Hay Derricks. Two styles of hay derricks are used to stack alfalfa hay in Idaho. The drawing to the left shows the one most in use because it is easier made and easier to move. The derrick to the right usually is made larger and more powerful. Wire cable is generally used with both derricks because rope wears out quickly. They are similar in operation but different in construction. The base of each is 16 feet square and the high ends of the booms reach up nearly 40 feet. A single hayfork rope, or wire cable, is used; it is about 65 feet long. The reach is sufficient to drop the hay in the center of a stack 24 feet wide.

127

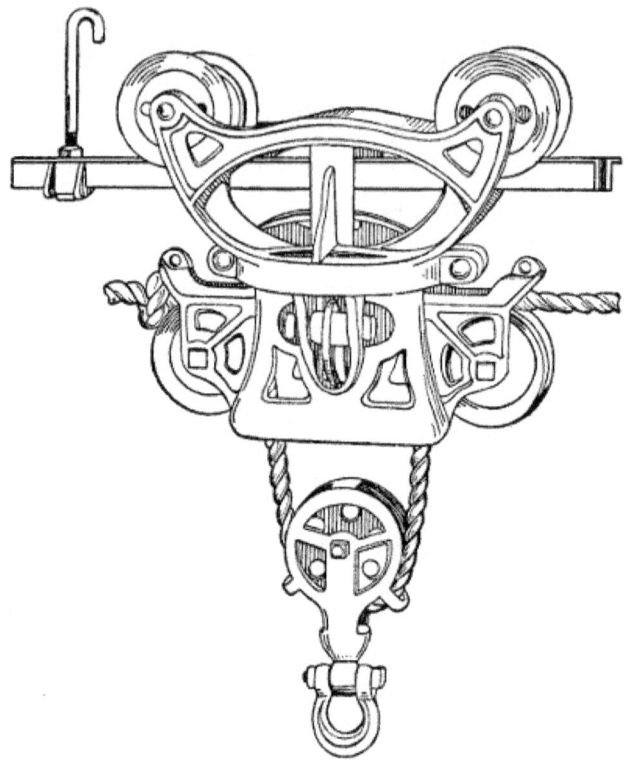

Figure 152.—Hay Carrier Carriage. Powerful carriers are part of the new barn. The track is double and the wheels run on both tracks to stand a side pull and to start quickly and run steadily when the clutch is released.

1   2

Figure 153.—(1) Hayfork Hitch. A whiffletree pulley doubles the speed of the fork. The knot in the rope gives double power to start the load. (2) Rafter Grapple, for attaching an extra pulley to any part of the barn roof.

There are timber braces fitted across the corners which are bolted through the outside timbers to brace the frame against a diamond tendency when moving the derrick. There is considerable strain when passing over uneven ground. It is better to make the frame so solid that it cannot get out of square. The mast is a stick of timber 8 inches square and 20 or 24 feet long. This mast is securely fastened solid to the center of the frame by having the bottom end mortised into the center cross timber at the middle and it is braced solid and held perpendicular to the framework by 4″ x 4″ wooden braces at the corners. These braces are notched at the top ends to fit the corners of the mast and are beveled at the bottom ends to fit flat on top of the timbers. They are held in place by bolts and by strap iron or band iron bands. These bands are drilled with holes and are spiked through into the timbers with four inch or five-inch wire nails. Holes are drilled through the band iron the right size and at the proper places for the nails. The mast is made round at the top and is fitted with a heavy welded iron ring or band to prevent splitting. The boom is usually about 30 feet long. Farmers prefer a round pole when they can get it. It is attached to the top of the mast by an iron stirrup made by a blacksmith. This stirrup is made to fit loosely half way around the boom one-third of the way up from the big end, which makes the small end of the boom project 20 feet out from the upper end of the mast. The iron stirrup is made heavy and strong. It has a round iron gudgeon 1½″ in diameter that reaches down into the top of the mast about 18 inches. The shoulder of the stirrup is supported by a square, flat iron plate which rests on and covers the top of the mast and has the corners turned down. It is made large to shed water and protect the top of the mast. This plate has a hole one and a half inches in diameter in the center through which the stirrup gudgeon passes as it enters the top of the mast. A farm chain, or logging chain, is fastened to the large end of the boom by passing the chain around the boom and engaging the round hook. The grab hook end of the chain is passed around the timber below and is hooked back to give it the right length, which doubles the part of the chain within reach of the man in charge. This double end of the chain is lengthened or shortened to elevate the outer end of the boom to fit the stack. The small outer end of the boom is thus raised as the stack goes up.

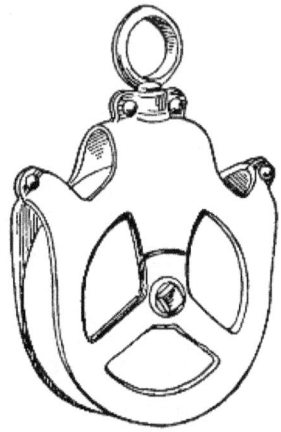
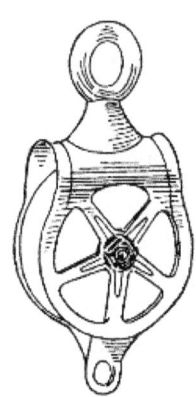

An ordinary horse fork and tackle is used to hoist the hay. Three single pulleys are attached, one to the outer end of the boom, one near the top of the mast, and the other at the bottom of the mast so that the rope passes easily and freely through the three pulleys and at the same time permits the boom to swing around as the fork goes up from the wagon rack over the stack. This swinging movement is regulated by tilting the derrick towards the stack so that the boom swings over the stack by its own weight or by the weight of the hay on the horse fork. Usually a wire truss is rigged over the boom to stiffen it. The wire is attached to the boom at both ends and the middle of the wire is sprung up to rest on a bridge placed over the stirrup.

Figure 155.—Gambrel Whiffletree, for use in hoisting hay to prevent entanglements. It is also handy when cultivating around fruit-trees.

Farmers like this simple form of hay derrick because it is cheaply made and it may be easily moved because it is not heavy. It is automatic and it is about as cheap as any good derrick and it is the most satisfactory for ordinary use. The base is large enough to make it solid and steady when in use. Before moving the point of the boom is lowered to a level position so that the derrick is not top-heavy. There is little danger of upsetting upon ordinary farm lands. Also the width of 16 feet will pass along country roads without meeting serious obstacles. Hay slings usually are made too narrow and too short. The ordinary little hay sling is prone to tip sideways and spill the hay. It is responsible for a great deal of profanity. The hay derrick shown to the right is somewhat different in construction, but is quite similar in action. The base is the same but the mast turns on a gudgeon stepped into an iron socket mortised into the center timber.

Figure 156.—Cable Hay Stacker. The wire cable is supported by the two bipods and is secured at each end by snubbing stakes. Two single-cable collars are clamped to the cable to prevent the bipods from slipping in at the top. Two double-cable clamps hold the ends of the cables to form stake loops.

The wire hoisting cable is threaded differently, as shown in the drawing. This style of derrick is made larger, sometimes the peak reaches up 40' above the base. The extra large ones are awkward to move but they build fine big stacks.

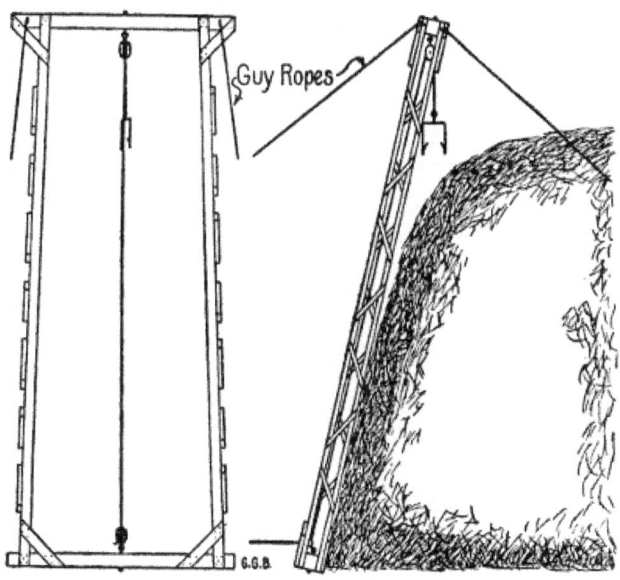

Figure 157.—California Hay Ricker, for putting either wild hay or alfalfa quickly in ricks. It is used in connection with home-made buck rakes. This ricker works against the end of the rick and is backed away each time to start a new bench. The upright is made of light poles or 2 x 4s braced as shown. It should be 28 or 30 feet high. Iron stakes hold the bottom, while guy wires steady the top.

CALIFORNIA HAY RICKER

In the West hay is often put up in long ricks instead of stacks. One of my jobs in California was to put up 2,700 acres of wild hay in the Sacramento Valley. I made four rickers and eight buck rakes similar to the ones shown in the illustrations. Each ricker was operated by a crew of eight men. Four men drove two buck rakes. There were two on the rick, one at the fork and one to drive the hoisting rig. Ten mowing machines did most of the cutting but I hired eight more machines towards the last, as the latest grass was getting too ripe. The crop measured more than 2,100 tons and it was all put in ricks, stacks and barns without a drop of rain on it. I should add that rain seldom falls in the lower Sacramento Valley during the haying season in the months of May and June. This refers to wild hay, which is made up of burr clover, wild oats and volunteer wheat and barley.

Alfalfa is cut from five to seven times in the hot interior valleys, so that if a farmer is rash enough to plant alfalfa under irrigation his haying thereafter will reach from one rainy season to the next.

CHAPTER VII

FARM CONVEYANCES

STONE-BOAT

One of the most useful and one of the least ornamental conveyances on a farm is the stone-boat. It is a low-down handy rig for moving heavy commodities in summer as well as in winter. No other sleigh or wagon will equal a stone-boat for carrying plows or harrows from one field to another. It is handy to tote bags of seed to supply the grain drill, to haul a barrel of water, feed for the hogs, and a great many other chores.

Figure 158.—Stone-Boat. Stump logs are selected for the planks. The bend of the planks is the natural curve of the large roots. The sawing is done by band saw cutting from two directions.

When the country was new, sawmills made a business of sawing stone-boat plank. Trees for stone-boat staves were cut close to the ground and the natural crooks of the roots were used for the noses of sleigh runners and for stone-boats. But cast-iron noses are now manufactured with recesses to receive the ends of straight ordinary hardwood planks. These cast-iron ends are rounded up in front to make the necessary nose crook. The front plank cross piece is bolted well towards the front ends of the runner planks. Usually there are two other hardwood plank cross pieces, one near the rear end and the other about one-third of the way back from the front. Placing the cross pieces in this way gives room between to stand a barrel.

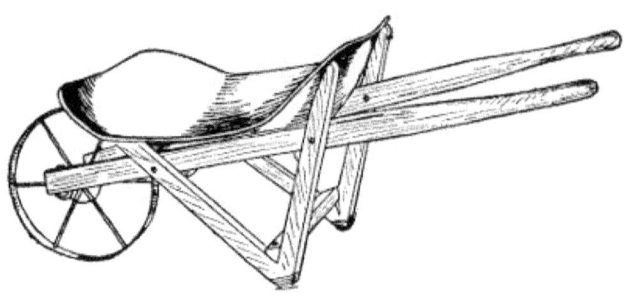

Figure 159.—Wheelbarrow. This factory-made wheelbarrow is the only pattern worth bothering with. It is cheap and answers the purpose better than the heavier ones with removable side wings.

The cross pieces are bolted through from the bottom up. Round-headed bolts are used and they are countersunk, to come flush with the bottom of the sliding planks. The nuts are countersunk into the cross pieces by boring holes about one-quarter inch deep. The holes are a little larger than the cornerwise diameter of the nuts. No washers are used, and the nuts are screwed down tight into the plank. The ends of the bolts are cut off even and filed smooth. The nuts are placed sharp corner side down and are left nearly flush on top or even with the surface of the cross pieces. In using a stone-boat, nobody wants a projection to catch any part of the load.

Regular doubletree clevises are attached to the corners of the old-fashioned stone-boat and the side chains are brought together to a ring and are just about long enough to form an equilateral triangle with the front end of the stone-boat. Cast-iron fronts usually have a projection in the center for the clevis hitch.

OXEN ON A NEW ENGLAND FARM

One of the most interesting experiences on a New England farm is to get acquainted with the manner in which oxen are pressed into farm service. One reason why oxen have never gone out of fashion in New England is the fact that they are patient enough to plow stony ground without smashing the plow.

A great deal of New England farm land has been reclaimed by removing a portion of the surface stone. In the processes of freezing and thawing and cultivation, stones from underneath keep working up to the surface so that it requires considerable skill to do the necessary plowing and cultivating. Oxen ease the plowpoint over or around a rock so it can immediately dip in again to the full depth of the furrow. A good yoke of cattle well trained

134

are gentle as well as strong and powerful.

Oxen are cheaper than horses to begin with and they are valuable for beef when they are not needed any longer as work animals. The Holstein breed seems to have the preference for oxen with New England farmers. The necessary harness for a pair of cattle consists of an ox yoke with a ringbolt through the center of the yoke, midway between the two oxen. A heavy iron ring about five inches in diameter, made of round iron, hangs from the ring bolt. There are two oxbows to hold the yoke in place on the necks of the cattle. A logging chain with a round hook on one end and a grab hook on the other end completes the yoking outfit.

The round hook of the chain is hitched into the ring in the plow clevis. The chain is passed through the large iron ring in the oxbow and is doubled back to get the right length. The grab hook is so constructed that it fits over one link of the chain flatwise so that the next link standing crosswise prevents it from slipping.

The mechanism of a logging chain is extremely simple, positive in action and especially well adapted to the use for which it is intended. The best mechanical inventions often pass without notice because of their simplicity. Farmers have used logging chains for generations with hooks made on this plan without realizing that they were profiting by a high grade invention that embodies superior merit.

In yoking oxen to a wagon the hitch is equally simple. The end of the wagon tongue is placed in the ring in the ox yoke, the round hook engages with a drawbolt under the hammer strap bar. The small grab hook is passed through the large yoke ring and is brought back and engaged with a chain link at the proper distance to stretch the chain taut.

The process of yoking oxen and hitching them to a wagon is one of the most interesting performances on a farm. The off ox works on the off side, or far side from the driver. He usually is the larger of the two and the more intelligent. The near (pronounced n-i-g-h) ox is nearest to the driver who walks to the left. Old plows turned the furrow to the right so the driver could walk on hard ground. In this way the awkwardness and ignorance of the near ox is played against the docility and superior intelligence of the off ox. In yoking the two together the yoke is first placed on the neck of the off ox and the near ox is invited to come under. This expression is so apt that a great many years ago it became a classic in the hands of able writers to suggest submission or slavery termed "coming under the yoke." Coming under the yoke, however, for the New England ox, in these days of abundant feeding, is no hardship. The oxen are large and powerful and the work they have to do is just about sufficient to give them the needed exercise to enjoy their alfalfa hay and feed of oats or corn.

TRAVOY

One of the first implements used by farm settlers in the timbered sections of the United States and Canada, was a three-cornered sled made from the fork of a tree. This rough sled, in the French speaking settlements, was called a "travoy." Whether it was of Indian or French invention is not known; probably both Indians and French settlers used travoys for moving logs in the woods before American history was much written. The legs or runners of a travoy are about five feet long. There is a bunk which extends crossways from one runner to the other, about half or two-thirds of the way back from the turned-up nose. This bunk is fastened to the runners by means of wooden pins and U-shaped bows fitted into grooves cut around the upper half of the bunk near the ends. Just back of the turned up nose is another cross piece in the shape of a stout wooden pin or iron bolt that is passed through an auger hole extending through both legs from side to side of the travoy. The underside of the crotch is hollowed out in front of the bolt to make room to pass the logging chain through so it comes out in front under the turned up nose.

Figure 160.—Travoy. A log-hauling sled made from the fork of a tree.

The front of the travoy is turned up, sled runner fashion, by hewing the wood with an axe to give it the proper shape. Travoys are used to haul logs from a thick woods to the skidways. The manner of using a travoy is interesting. It is hauled by a yoke of cattle or a team of horses to the place where the log lies in the woods. The round hook end of the logging chain is thrown over the butt end of the log and pulled back under the log then around the bunk just inside of the runner and hooked fast upon itself. The travoy is then leaned over against the log, the grab hook end of the chain is brought over the log and over the travoy and straightened out at right angles to the log. The cattle are hitched to the end of the logging chain and started. This kind of a hitch rolls the log over on top of the bunk on the travoy. The cattle are then unhitched. The grab hook end of the chain thus released is passed down and around under the other end of the bunk from behind. The chain is then passed over the bolt near the nose of the travoy and pulled down through the opening and out in front from under the nose. The small grab hook of the logging chain is then passed through the clevis, in the doubletree, if horses are used, or the ring in the yoke if cattle are used, and hitched back to the proper length. A little experience is necessary to regulate the length of the chain to give the proper pull. The chain should be short enough so the pull lifts a little. It is generally conceded by woodsmen that a short hitch moves a log easier than a long hitch. However, there is a medium. There are limitations which experience only can determine. A travoy is useful in dense woods where there is a good deal of undergrowth or where there are places so rough that bobsleighs cannot be used to advantage.

LINCHPIN FARM WAGONS

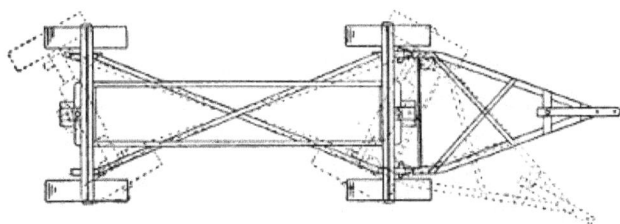

Figure 161.—Cross Reach Wagon. This wagon is coupled for a trailer, but it works just as well when used with a tongue and horses as a handy farm wagon. The bunks are made rigid and parallel by means of a double reach. There are two king bolts to permit both axles to turn. Either end is front.

Figure 162.—Wagon Brake. The hounds are tilted up to show the brake beam and the

137

manner of attaching it. The brake lever is fastened to the forward side of the rear bolster and turns up alongside of the bolster stake. The brake rod reaches from the upper end of the lever elbow to the foot ratchet at the front end of the wagon box.

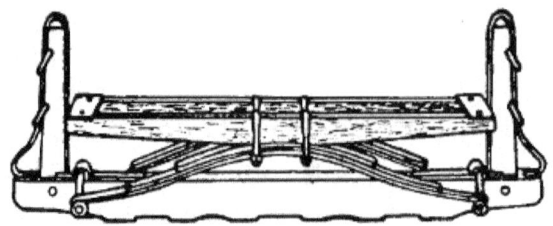

Figure 163.—Bolster Spring.

In some parts of the country the wheels of handy wagons about the farm are held on axle journals by means of linchpins in the old-fashioned manner. There are iron hub-bands on both ends of the hubs which project several inches beyond the wood. This is the best protection against sand to prevent it from working into the wheel boxing that has ever been invented. Sand from the felloes scatters down onto these iron bands and rolls off to the ground. There is a hole through each band on the outer ends of the hubs to pass the linchpin through so that before taking off a wheel to oil the journal it must first be turned so the hole comes directly over the linchpin. To pry out the linchpin the drawbolt is used. Old-fashioned drawbolts were made with a chisel shaped end tapered from both sides to a thickness of about an eighth of an inch. This thin wedge end of the drawbolt is placed under the end of the linchpin. The lower side of the hub-band forms a fulcrum to pry the pin up through the hole in the upper side of the sand-band projection. The linchpin has a hook on the outer side of the upper end so the lever is transferred to the top of the sand-band when another pry lifts the pin clear out of the hole in the end of the axle so the wheel may be removed and grease applied to the axle. The drawbolt on a linchpin wagon usually has a head made in the form of the jaws of a wrench. The wrench is the right size to fit the nuts on the wagon brace irons so that the drawbolt answers three purposes.

Figure 164.—Wagon Seat Spring. The metal block fits over the top of the bolster stake.

Figure 165.—Hollow Malleable Iron Bolster Stake to hold a higher wooden stake when necessary.

SAND-BANDS

Many parts of farm machinery require projecting sand-bands to protect the journals from sand and dust. Most farms have some sandy fields or ridges. Some farms are all sand or sandy loam. Even dust from clay is injurious to machinery. There is more or less grit in the finest clay. The most important parts of farm machinery are supposed to be protected by oil-cups containing cotton waste to strain the oil, together with covers in the shape of metal caps. These are necessary protections and they help, but they are not adequate for all conditions. It is not easy to keep sand out of bearings on machinery that shakes a good deal. Wooden plugs gather sand and dust. When a plug is pulled the sand drops into the oil hole. Farm machinery that is properly designed protects itself from sand and dust. In buying a machine this particular feature should appeal to the farmers more than it does. Leather caps are a nuisance. They are a sort of patchwork to finish the job that the manufacturer commences. A man who is provident enough to supply himself with good working tools and is sufficiently careful to take care of them, usually is particular about the appearance as well as the usefulness of his tools, machinery and implements.

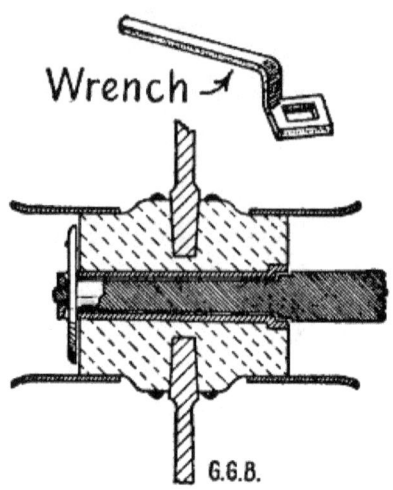

Figure 166.—Sand Caps. Not one manufacturer in a hundred knows how to keep sand out of an axle bearing. Still it is one of the simplest tricks in mechanics. The only protection an axle needs is long ferrules that reach out three or four inches beyond the hub at both ends. Old-fashioned Linchpin farm wagons were built on this principle. The hubs held narrow rings instead of skeins, but they wore for years.

BOBSLEIGHS

On Northern farms bobsleighs are as important in the winter time as a farm wagon in summer. There are different ways of putting bobsleighs together according to the use required of them. When using heavy bobsleighs for road work, farmers favor the bolster reach to connect the front and rear sleighs. With this attachment the horses may be turned around against the rear sled. The front bolster fits into a recessed plate bolted to the bench plank of the front sleigh. This plate is a combination of wearing plate and circle and must be kept oiled to turn easily under a heavy load. It not only facilitates turning, but it prevents the bolster from catching on the raves or on the upturned nose of the front bob when turning short.

The heavy hardwood plank reach that connects the two bolsters is put through a mortise through the front bolster and is fastened rigidly by an extra large king-bolt. The reach plays back and forth rather loosely through a similar mortise in the other bolster on the rear sleigh. The rear hounds connect with the reach by means of a link and pin. This link pushes up through mortise holes in the reach and is fastened with a wooden pin or key on top of the reach. Sometimes the hounds are taken away and the reach is fastened with pins before and behind the rear bolster. This reach hitch is not recommended except for light road work. These two ways of attaching the rear sled necessitate different ways of fastening the rear bolster to the sled. When the rear bolster is required to do the pulling, it is attached to the sled by double eyebolts which permit the necessary rocking motion and allows the nose of the rear sled to bob up and down freely. This is an advantage when a long box bed is used, because the bolster is made to fit the box closely and is not continually oscillating and wearing. Eye-bolts provide for this natural movement of the sled. Light pleasure bobs are attached to the box with eyebolts without bolster stakes. The light passenger riding seat box is bound together with iron braces and side irons so it does not need bolsters to hold the sides together.

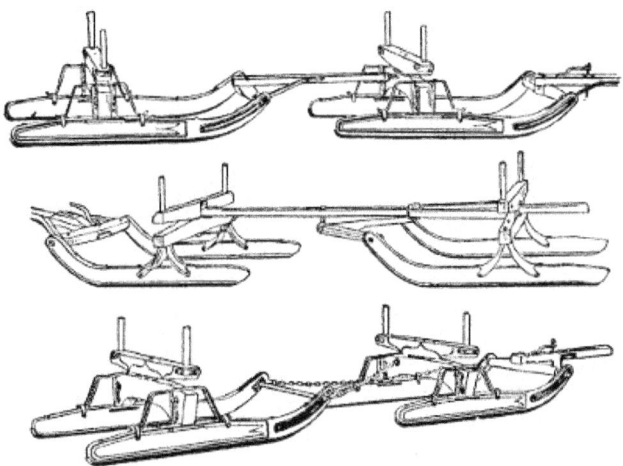

Figure 167.—Bobsleighs, Showing Three Kinds of Coupling. The upper sleighs are coupled on the old-fashioned short reach plan except that the reach is not mortised into the roller. It is gained in a quarter of an inch and fastened by an iron strap with a plate and nuts on the under side. The bobs in the center show the bolster reach, principally used for road work. The bottom pair are coupled by cross chains for short turning around trees and stumps in the woods.

Bobsleighs for use in the woods are hitched together quite differently. The old-fashioned reach with a staple in the rear bench of the first sled and a clevis in the end of the reach is the old-fashioned rig for rough roads in the woods. Such sleighs are fitted with bunks instead of bolsters. Bunks are usually cut from good hardwood trees, hewed out with an axe and bored for round stakes. Log bunks for easy loading do not project beyond the raves. With this kind of a rig, a farmer can fasten two logging chains to the reach, carry the grab hook ends out and under and around the log and back again over the sleighs, and then hitch the horses to the two chains and roll the log up over a couple of skids and on to the bunks without doing any damage to the bobsleighs. Bobsleighs hitched together with an old-

141

fashioned reach and provided with wide heavy raves will climb over logs, pitch down into root holes, and weave their way in and out among trees better than any other sled contrivance, and they turn short enough for such roads. The shortest turning rig, however, is the cross chain reach shown in Figure 167.

MAKING A FARM CART

A two-wheeled cart large enough to carry a barrel of cider is a great convenience on a farm. The front wheels of a buggy are about the right size and usually are strong enough for cart purposes. A one-inch iron axle will be stiff enough if it is reinforced at the square bends. The axle is bent down near the hubs at right angles and carried across to support the floor of the cart box about one foot from the ground. The distance from the ground should be just sufficient so that when the cart is tipped back the hind end will rest on the ground with the bottom boards at an easy slant to roll a barrel or milk can into the bottom of the box. Under the back end of the cart platform is a good stout bar of hardwood framed into the sidepieces. All of the woodwork about the cart is well braced with iron. The floor of the cart is better when made of narrow matched hardwood flooring about seven-eighths of an inch thick fastened with bolts. It should be well supported by cross pieces underneath. In fact the principal part of the box is the underneath part of the frame.

Sidepieces of the box are wide and are bolted to the vertical parts of the axle and braced in different directions to keep the frame solid, square and firm. The sides of the box are permanently fastened but both tailboard and front board are held in place by cleats and rods and are removable so that long scantling or lumber may be carried on the cart bottom. The ends of the box may be quickly put in place again when it is necessary to use them.

To hold a cart box together, four rods are necessary, two across the front and two behind. They are made like tailboard rods in wagon boxes. There is always some kind of tongue or handle bar in front of the farm cart conveniently arranged for either pulling or pushing. If a breast bar is used it handles better when supported by two curved projecting shafts or pieces of bent wood, preferably the bent up extended ends of the bedpieces. The handle bar should be about three feet from the ground.

Figure 168.—Farm Cart. The axle need not be heavier than ⅞". The hind axle of a light buggy works the best. It is bent down and spliced and welded under the box. The cart should be made narrow to prevent overloading. The box should be low enough to rest the back end on the ground at an angle of about 35° for easy loading.

COLT-BREAKING SULKY

A pair of shafts that look a good deal too long, an axle, two wheels and a whiffletree are the principal parts of a colt-breaking sulky. The shafts are so long that a colt can kick his best without reaching anything behind. The principal danger is that he may come down with one hind leg over the shaft. It is a question with horsemen whether it is better to first start a colt alongside of an old, steady horse. But it is generally conceded that in no case should a colt be made fast in such a way that he could kick himself loose. Different farmers have different ideas in regard to training colts, but these breaking carts with extra long shafts are very much used in some parts of the country. The shafts are heavy enough so that the colts may be tied down to make kicking impossible. A rope or heavy strap reaching from one shaft to the other over the colt's hips will keep its hind feet pretty close to the ground. Any rig used in connection with a colt should be strong enough to withstand any strain that the colt may decide to put upon it. If the colt breaks something or breaks loose, it takes him a long time to forget the scare. Farm boys make these breaking carts by using wheels and hind axles of a worn-out buggy. This is well enough if the wheels are strong and shafts thoroughly bolted and braced. It is easy to make a mistake with a colt. To prevent accidents it is much better to have the harness and wagon amply strong.

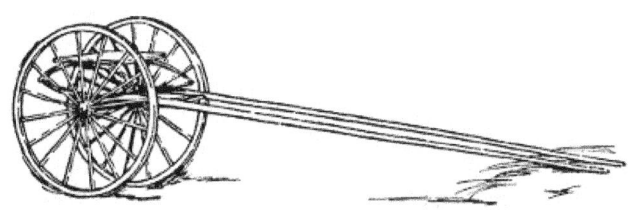

Figure 169.—Colt-Breaking Sulky. The axle and hind wheels of a light wagon, two strong straight-grained shafts about 4 feet too long, a whiffletree and a spring seat are the principal parts of a colt-breaking sulky. The shafts and seat are thoroughly well bolted and clipped to the axle and braced against all possible maneuvers of the colt. The traces are made so long that the colt cannot reach anything to kick, and he is prevented from kicking by a strap reaching from one shaft up over his hips and down to the other shaft. In this rig the colt is compelled to go ahead because he cannot turn around. The axle should be longer than standard to prevent upsetting when the colt turns a corner at high speed.

CHAPTER VIII

MISCELLANEOUS FARM CONVENIENCES

FARM OFFICE

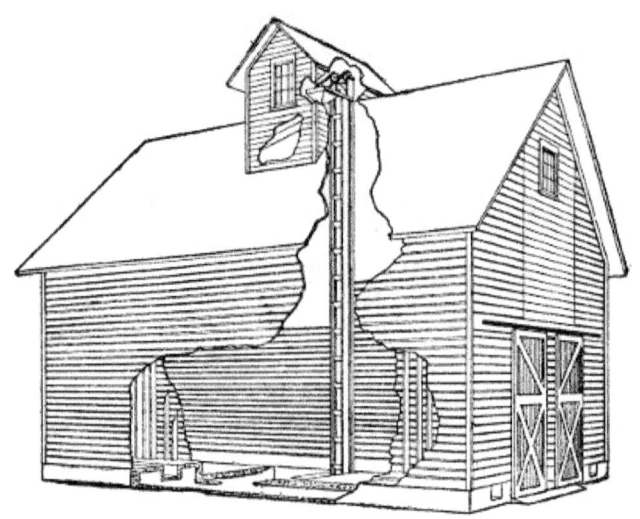

Figure 170.—Perspective View of Two-Story Corn Crib. The side of the building is cut away to show the elevating machinery.

Business farming requires an office. Business callers feel sensitive about talking farm or live-stock affairs before several members of the family. But they are quite at ease when alone with the farmer in his office. A farm office may be small but it should contain a desk or table, two or three chairs, book shelves for books, drawers for government bulletins and a cabinet to hold glassware and chemicals for making soil tests and a good magnifying glass for examining seeds before planting. A good glass is also valuable in tracing the destructive work of many kinds of insect pests.

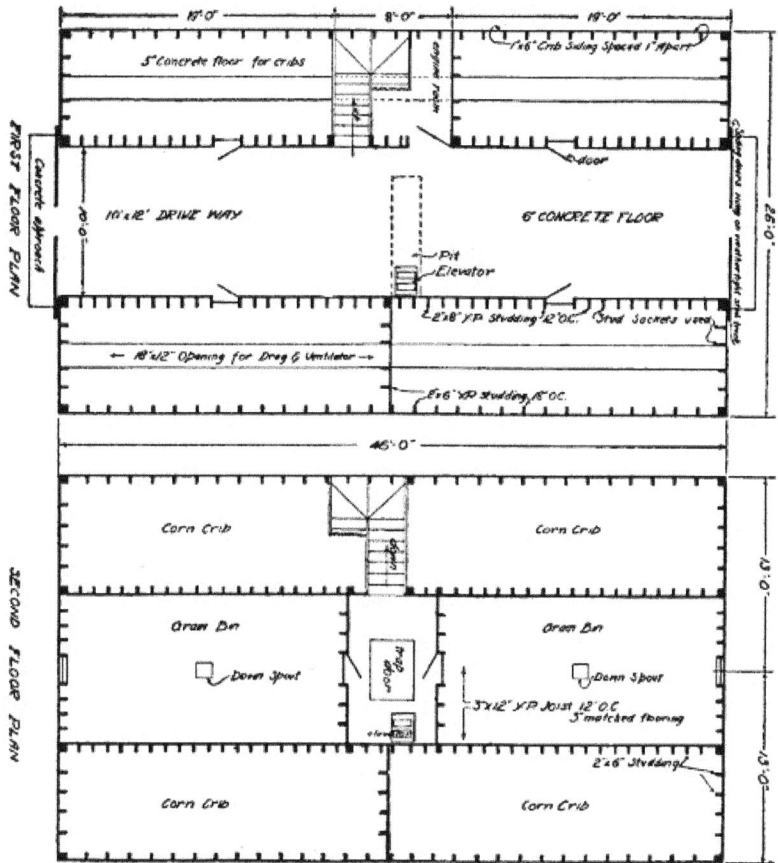

Figure 171.—Floor Plans of Two-Story Corn Crib. The first floor shows the driveway with corn cribs at the sides and the second floor plan shows the grain bins over the center driveway, with location of the downspouts, stairway, etc.

Large scale image (1383 x 1500, 66 kB)

The office is the proper place for making germination tests of various farm seeds. Seventy degrees of heat is necessary for the best results in seed testing. For this reason, as well as for comfort while working, the heating problem should receive its share of attention. Many times it so happens that a farmer has a few minutes just before mealtime that he could devote to office work if the room be warm enough.

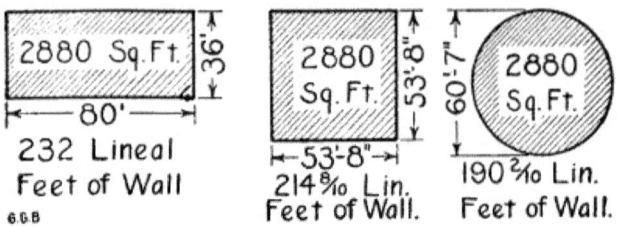

Figure 172.—Economy of Round Barn. The diagrams show that the popular 36' x 80' cow stable and the commonest size of round barn have about the same capacity. Each barn will stable forty cows, but the round barn has room for a silo in the center. Both barns have feed overhead in the shape of hay and straw, but the round feed room saves steps.

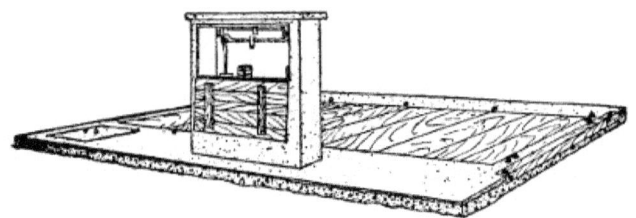

Figure 173.—Concrete Farm Scale Base and Pit.

Neatly printed letter-heads and envelopes are important. The sheets of paper should be eight and a half by eleven inches in size, pure white and of good quality. The printing should be plain black and of round medium-sized letters that may be easily read. Fancy lettering and flourishes are out of place on business stationery.

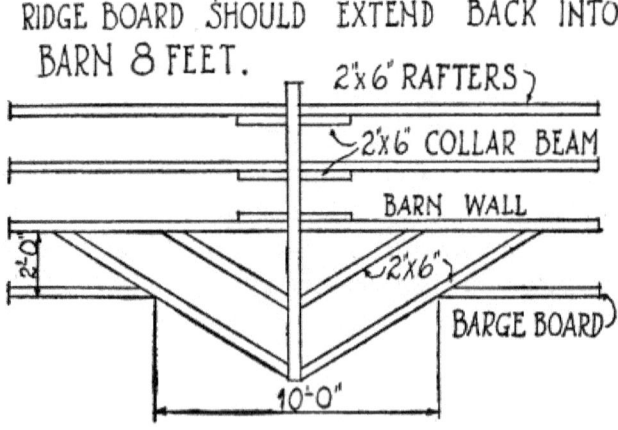

Figure 174.—Top View of the Hay-Track Roof Extension, showing the ridgeboard and supporting jack-rafters.

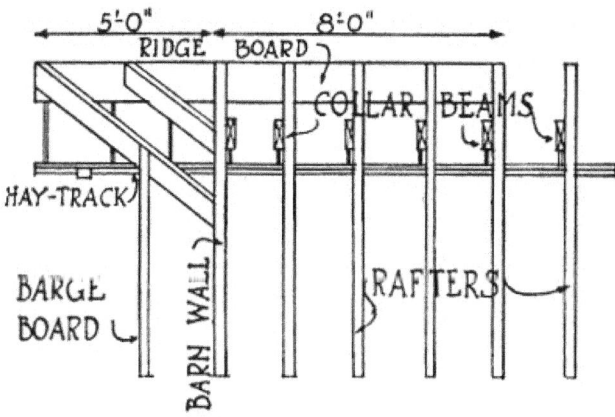

Figure 175.—Side view showing plan for building a Hayfork Hood to project from peak of a storage barn. The jack-rafters form a brace to support the end of the hay-track beam.

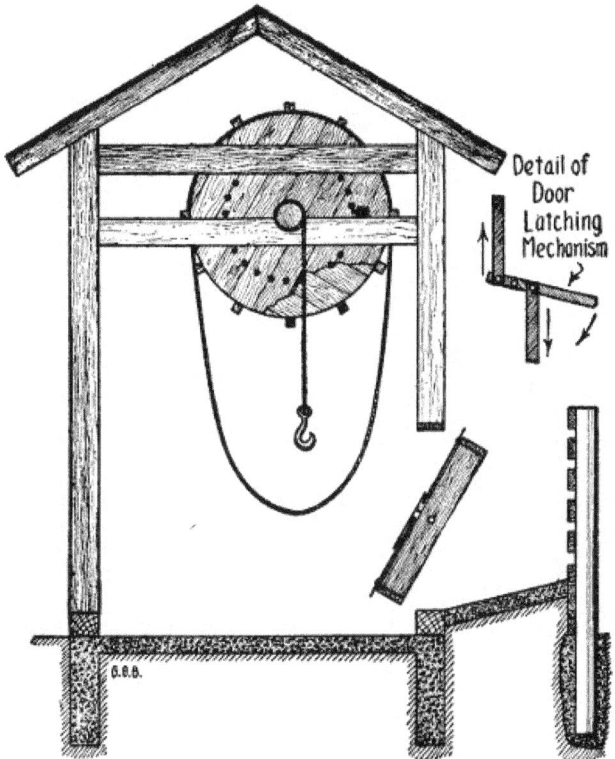

Figure 176.—Slaughter House. The house should be twelve feet wide. It may be any length to provide storage, but 12 x 12 makes a good beef skinning floor. The windlass shaft should be ten feet above the floor, which requires twelve-foot studding. The wheel is eight feet in diameter and the winding drum is about ten inches. The animal is killed on the incline outside of the building and it lies limp against the revolving door. The door catch is sprung back and the carcass rolls down onto the concrete skinning floor.

Halftone illustration of farm animals or buildings are better used on separate advertising sheets that may be folded in with the letters when wanted.

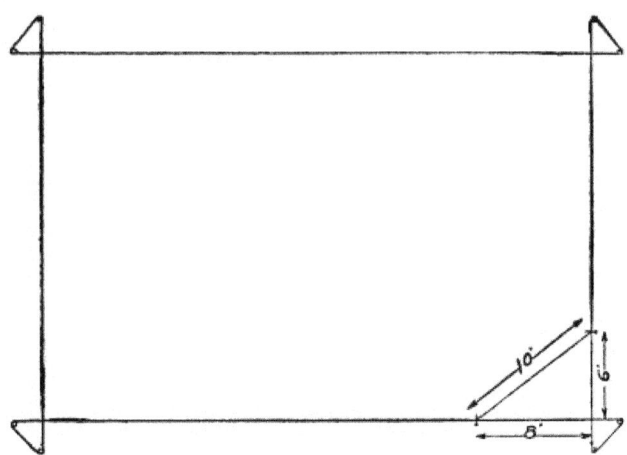

Figure 177.—Rule of Six, Eight and Ten. Diagram showing how to stake the foundation of a farm building so the excavation can be made clear out to the corners without undermining the stakes.

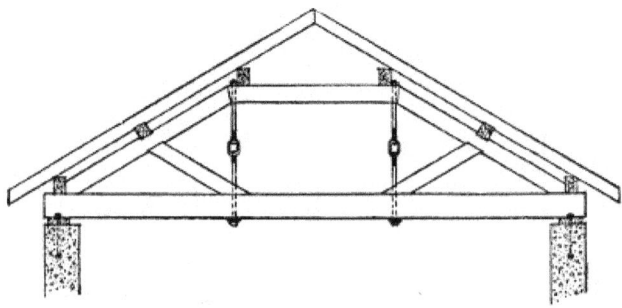

Figure 178.—Roof Truss built strong enough to support the roof of a farm garage without center posts.

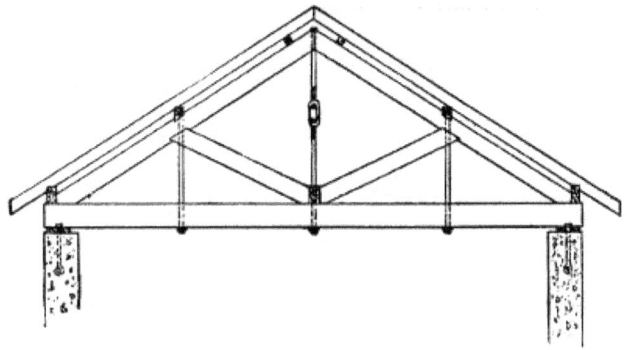

Figure 179.—Design of Roof Truss Intended to Span a Farm Garage.

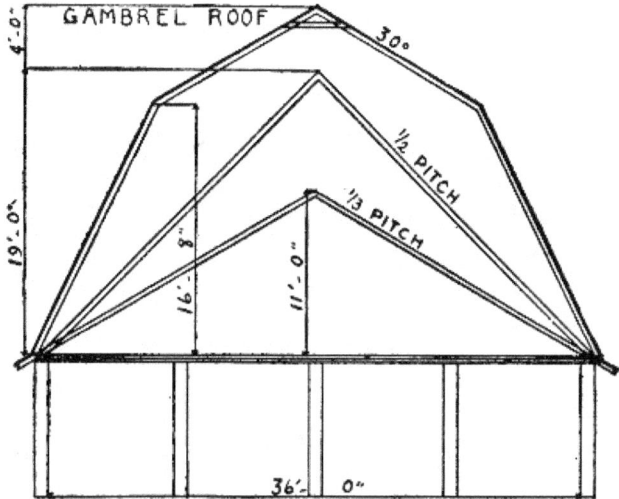

Figure 180.—Roof Pitches. Mow capacity of the different roof pitches is given above the plates in figures.

Typewriters are so common that a hand-written letter is seldom seen among business correspondence. A busy farmer is not likely to acquire much speed with a typewriter, but his son or daughter may. One great advantage is the making of carbon copies. Every letter received is then filed in a letter case in alphabetical order and a carbon copy of each answer is pinned to it for future reference.

151

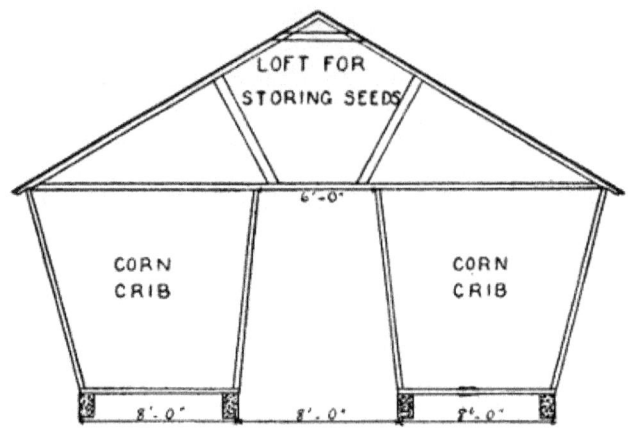

Figure 181.—Double Corn Crib. Two cribs may be roofed this way as cheaply as to roof the two cribs separately. A storeroom is provided overhead and the bracing prevents the cribs from sagging.

The cost of furnishing a farm office will depend upon the inclinations of the man. A cheap kitchen table may be used instead of an expensive mahogany desk. A new typewriter costs from fifty to ninety dollars, but a rebuilt machine that will do good work may be obtained for twenty.

A useful magnifying glass with legs may be bought for a dollar or two. Or considerable money may be invested in a high-powered microscope.

SPEED INDICATOR

The speed requirements of machines are given by the manufacturers. It is up to the farmer to determine the size of pulleys and the speed of intermediate shafts between his engine and the machine to be driven. A speed indicator is held against the end of a shaft at the center. The indicator pin then revolves with the shaft and the number of revolutions per minute are counted by timing the pointer on the dial with the second hand of a watch.

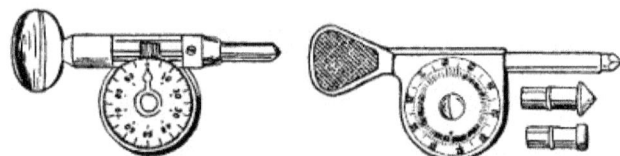

Figure 182.—Speed Timers. Two styles. The point is held against the center of the shaft to be tested. The number of revolutions per minute is shown in figures on the face of the dial. The indicator is timed to the second hand of a watch.

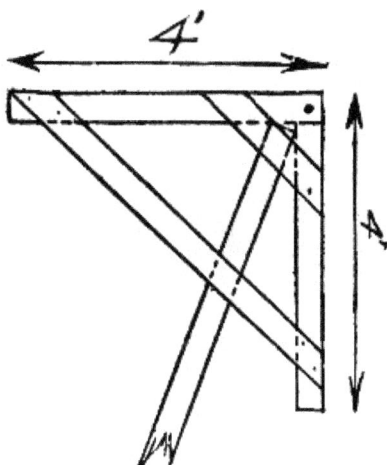

Figure 183.—Building Bracket. Made of 2 x 4 pieces put together at right angles with diagonal braces. The supporting leg fits between the four diagonal braces.

SOIL TOOLS

Soil moisture often is the limiting factor in crop raising. Soil moisture may be measured by analysis.[203-204-205] The first step is to obtain samples at different depths. This is done accurately and quickly with a good soil auger. Other paraphernalia is required to make a careful analysis of the sample, but a farmer of experience will make a mud ball and form a very good estimate of the amount of water in it.

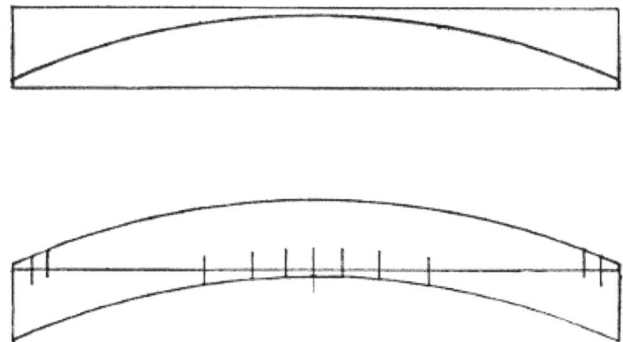

Figure 184.—Diagram showing how to cut a plank on a band-saw to form a curved rafter. The two pieces of the plank are spiked together as shown in the lower drawing. This makes a curved rafter without waste of material.

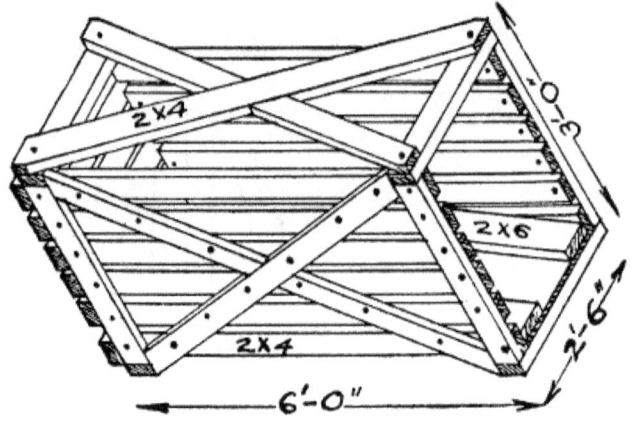

Figure 185.—Breeding Crate for Hogs. The illustration shows the manner of construction.

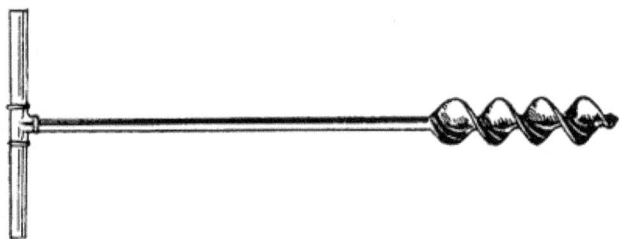

Figure 186.—Soil Auger. Scientific farming demands that soils shall be tested for moisture. A long handled auger is used to bring samples of soil to the surface. The samples are weighed, the water evaporated and the soil reweighed to determine the amount of moisture.

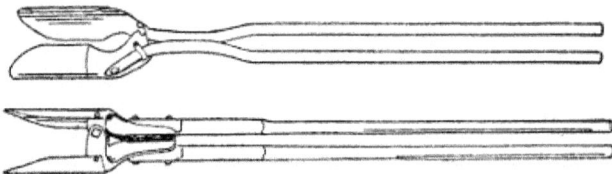

Figure 187.—Post Hole Diggers. Two patterns of the same kind of digger are shown. The first has iron handles, the lower has wooden handles.

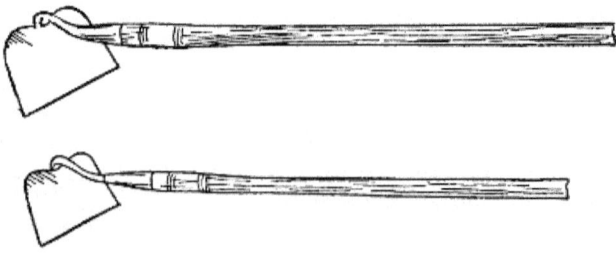

Figure 188.—Hoes and Weeders. The hang of a hoe affects its working. The upper hoe shows about the easiest working angle between the blade and the handle. The difference between a hoe and a weeder is that the hoe is intended to strike into the ground to loosen the soil, while the blade of the weeder is intended to work parallel with the surface of the soil to cut young weeds.

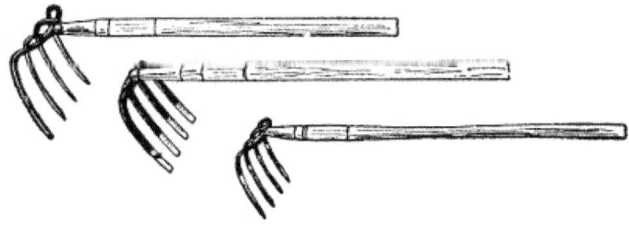

Figure 189.—Manure Hook and Potato Diggers.

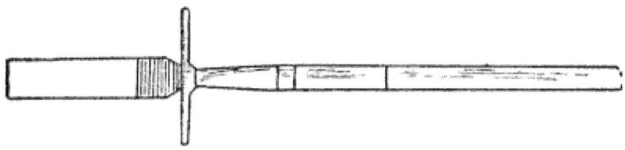

Figure 190.—Spud. Certain vegetables are grown for crop and for seed. The green plants are thinned with a spud for sale, leaving the best to ripen for seed. It is also used to dig tough weeds, especially those having tap roots.

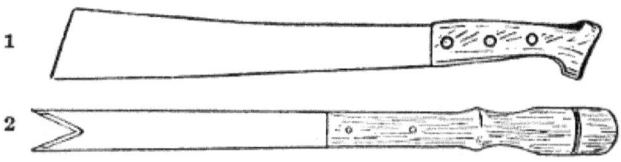

Figure 191.—(1) Corn Cutting Knife. (2) Asparagus Knife.

FENCE-MAKING TOOLS

Sliding Field Gate.—Each farm field should have a gate, not necessarily expensive, but it should be reasonably convenient. Farm field gates should be made sixteen feet long, which will allow for a clear opening about fourteen feet wide. The cheapest way to make a good farm gate is to use a 10-inch board for the bottom, 8-inch for the board next to the bottom and three 6-inch boards above that. The space between the bottom board and next board is two inches. This narrow space prevents hogs from lifting the gate with their noses. The spaces widen toward the top, so that the gate when finished is five feet high. If colts run the fields then a bar is needed

155

along the top of the gate. Six cross pieces 1 inch by 6 inches are used to hold the gate together. These cross pieces are bolted through at each intersection. Also a slanting brace is used on the front half of the gate to keep it from racking and this brace is put on with bolts. Two posts are set at each end of the gate. The front posts hold the front end of the gate between them, and the rear posts the same. There is a cross piece which reaches from one of the rear posts to the other to slide the gate and hold it off the ground. A similar cross piece holds the front end of the gate up from the ground. Sometimes a swivel roller is attached to the rear cross piece to roll the gate if it is to be used a good deal. A plain, simple sliding gate is all that is necessary for fields some distance from the barn.

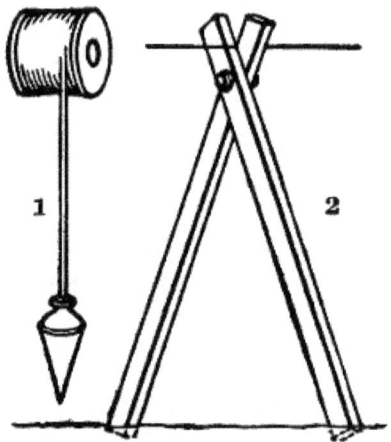

Figure 192.—(1) Plumb-Bob and Plumb-Line. The line is paid out about 6 feet from the spool and given a half hitch. It may then be hung over the wire and the spool will balance the bob. (2) Bipod. The legs of a fence bipod are cut 6 feet long. The bolt is put through 6 inches from the top ends. By the aid of the plummet the upper wire is strung plumb over the barb-wire in the furrow and 4' 6" above grade. The lower parts of the posts are set against the barb-wire and the upper faces of the posts at the top are set even with the upper wire. This plan not only places the posts in line, both at the top and bottom, but it regulates the height.

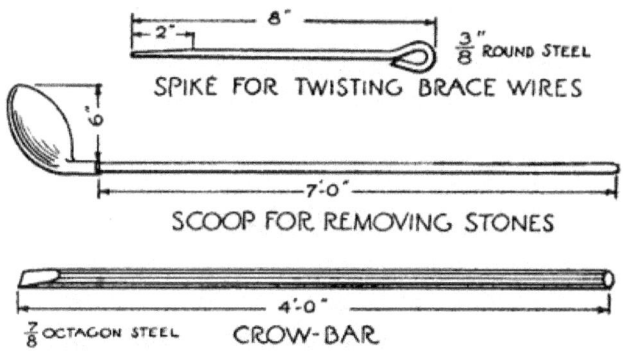

Figure 193.—Fence Tools. The upper tool is a round steel pin to twist heavy brace wires. The scoop is for working stones out of post-holes. The steel crowbar is for working around the stones to loosen them.

Figure 194.—Fence Pliers. This is a heavy fence tool made to pull fence staples and to stretch, cut and splice wire.

CORN SHOCK HORSE

Figure 195.—Corn Horse. When corn is cut by hand there is no better shocking device than the old-style corn horse. It is almost as handy when setting up the corn sheaves from the corn binder.

A convenient corn shocking horse is made with a pole cut from a straight tree. The pole is about six inches through at the butt and tapers to a small end. About twenty feet is a good length. There are two legs which hold the large end of the pole up about 40" from the ground. These legs are well spread apart at the bottom. Two feet back from the legs is a horizontal hole about one and one-quarter inches in diameter to hold the crossbar. This crossbar may be an old broom handle. The pole and the crossbar mark the four divisions of a corn shock. Corn is cut and stood up in each corner, usually nine hills in a corner, giving thirty-six hills to a shock. Corn planted in rows is counted up to make about the same amount of corn to the shock. Of course a heavy or light crop must determine the number of rows or hills. When enough corn is cut for a shock it is tied with two bands, the crossbar is pulled out and the corn horse is dragged along to the next stand.

HUSKING-PIN

Hand huskers for dividing the cornhusks at the tips of the ears are made of wood, bone or steel. Wooden husking-pins are made of ironwood, eucalyptus, second growth hickory, or some other tough hardwood. The[209-210-211]

pin is about four inches long, five-eighths of an inch thick and it is shaped like a lead-pencil with a rather long point. A recessed girdle is cut around the barrel of the pin and a leather finger ring fits into and around this girdle. Generally the leather ring fits the larger finger to hold the pin in the right position while permitting it to turn to wear the point all around alike. Bone husking-pins are generally flat with a hole through the center to hold the leather finger ring. Steel husking-pins are shaped differently and have teeth to catch and tear the husks apart.

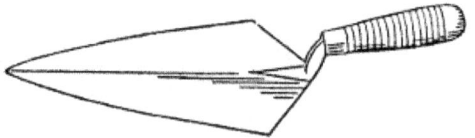

Figure 196.—Brick Trowel.

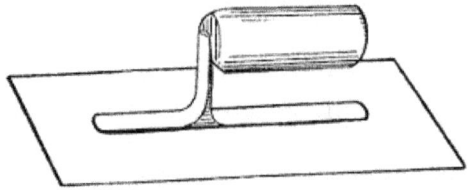

Figure 197.—Plastering Trowel.

158

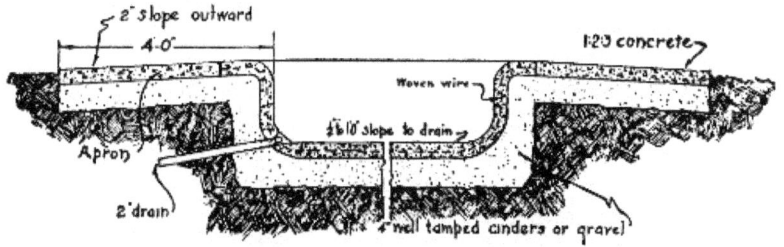

Figure 198.—Concrete Hog Wallow, showing drain pipe.

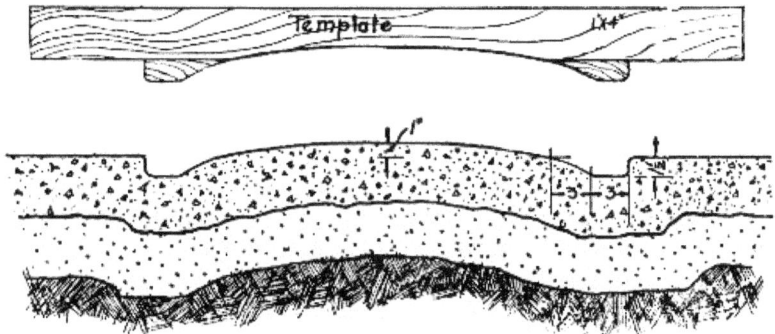

Figure 199.—Concrete Center Alley for Hog House. The upper illustration represents the wooden template used to form the center of the hog house floor.

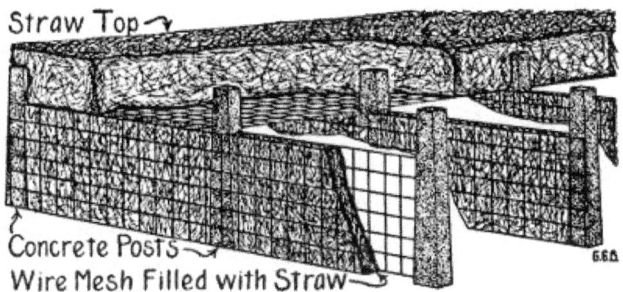

Figure 200.—Sanitary Pig-Pen. One of the most satisfactory farrowing houses is constructed of concrete posts 6″ square and 6″ square mesh hog fencing and straw. The posts are set to make farrowing pens 8′ wide and 16′ deep from front to back. Woven wire is stretched and fastened to both sides of the posts at the sides and back of each pen. Straw is stuffed in between the two wire nets, thus making partitions of straw 6″ thick and 42″ high. Fence wire is stretched over the top and straw piled on deep enough to shed rain. The front of the pens face the south and are closed by wooden gates. In the spring the pigs are turned out on pasture, the straw roof is hauled to the fields for manure and the straw partitions burned out. The sun shines into the skeleton pens all summer so that all mischievous bacteria are killed and the hog-lice are burned or starved. The next fall concrete floors may be laid in the pens, the partitions restuffed with straw and covered with another straw roof. In a colder climate I would cover the whole top with a straw roof. Sufficient ventilation would work through the straw partitions and the front gate. In very cold weather add a thin layer of straw to the gate.

159

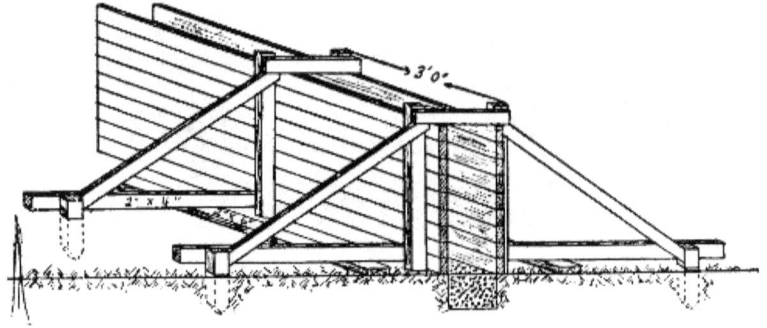

Figure 201.—Concrete Wall Mold. Wooden molds for shaping a concrete wall may be made as shown. If the wall is to be low—2′ or less—the mold will stay in place without bolting or wiring the sides together. The form is made level by first leveling the 2″ x 6″ stringers that support the form.

Figure 202.—Husking-Pin. The leather finger ring is looped into the recess in the wooden pin.

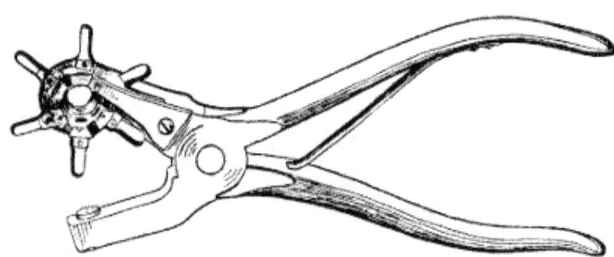

Figure 203.—Harness Punch. The hollow punch points are of different sizes.

Figure 204.—Belt Punch. Two or three sizes should be kept in the tool box. Belt holes should be small to hold the lace tight. The smooth running of belts depends a good deal on the lacing. Holes punch better against the end of a hickory block or other fine grained hardwood.

PAINT BRUSHES

Paint brushes may be left in the paint for a year without apparent injury. The paint should be deep enough to nearly bury the bristles. Pour a little

160

boiled linseed oil over the top to form a skin to keep the air out. It is cheaper to buy a new brush than to clean the paint out of one that has been used.

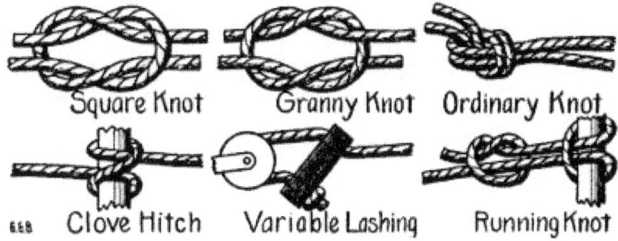

Figure 205.—Knots. The simple principles of knot tying as practiced on farms are here represented.

Figure 206.—Sheepshank, two half hitches in a rope to take up slack. The rope may be folded upon itself as many times as necessary.

Figure 207.—Marline Spike. Used for splicing ropes, tying rose knots, etc.

FRUIT PICKING

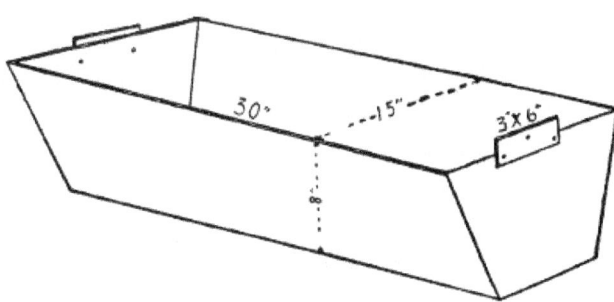

Figure 208.—Fruit-Picking Tray. It is used for picking grapes and other fruits. The California lug box has vertical sides and is the same size top and bottom. Otherwise the construction is similar.

161

Apples are handled as carefully as eggs by men who understand the business of getting high prices. Picking boxes for apples have bothered orchard men more than any other part of the business. It is so difficult to get help to handle apples without bruising that many inventions have been tried to lessen the damage. In western New York a tray with vertical ends and slanting sides has been adopted by grape growers as the most convenient tray for grapes. Apple growers are adopting the same tray. It is made of three-eighths-inch lumber cut 30 inches long for the sides, using two strips for each side. The bottom is 30 inches long and three-eighths of an inch thick, made in one piece. The ends are seven-eighths of an inch thick cut to a bevel so the top edge of the end piece is fourteen inches long and the bottom edge is ten inches long. The depth of the end piece is eight inches. Hand cleats are nailed on the outsides of the end pieces so as to project one-half inch above the top. These cleats not only serve to lift and carry the trays, but when they are loaded on a wagon the bottoms fit in between the cleats to hold them from slipping endways. In piling these picking boxes empty, one end is slipped outward over the cleat until the other end drops down. This permits half nesting when the boxes are piled up for storage or when loaded on wagons to move to the orchard.

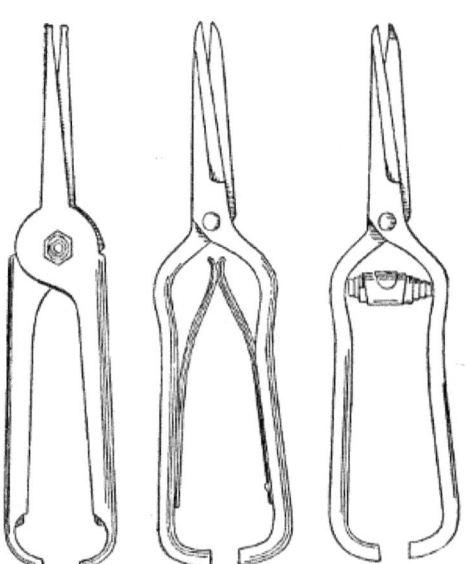

Figure 209.—Fruit Thinning Nippers. Three styles of apple-stem cutters are shown. They are also used for picking grapes and other fruits.

Apples are picked into the trays from the trees. The trays are loaded on to wagons or stone-boats and hauled to the packing shed, where the apples are rolled out gently over the sloping sides of the crates on to the cushioned bottom of the sorting table. Orchard men should have crates enough to keep the pickers busy without emptying until they are hauled to the packing shed. The use of such trays or crates save handling the apples over several times. The less apples are handled the fewer bruises are made.

Figure 210.—Apple Picking Ladder. When apples are picked and placed in bushel trays a ladder on wheels with shelves is convenient for holding the trays.

In California similar trays are used, but they have straight sides and are called lug boxes. Eastern fruit men prefer the sloping sides because they may be emptied easily, quickly and gently.

FRUIT PICKING LADDERS

Commercial orchards are pruned to keep the bearing fruit spurs as near the ground as possible, so that ladders used at picking time are not so long as they used to be.

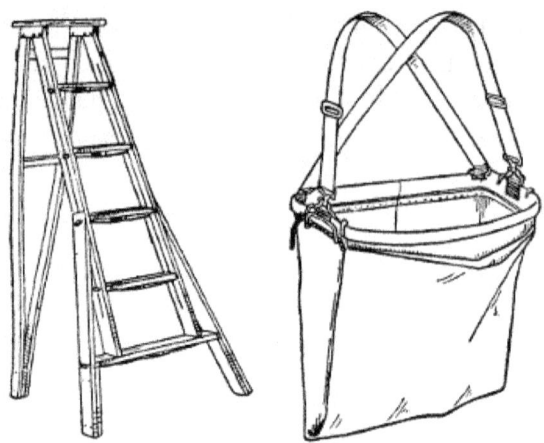

Figure 211.—Stepladder and Apple-Picking Bag. This ladder has only three feet, but the bottom of the ladder is made wide to prevent upsetting. This bag is useful when picking scattering apples on the outer or upper branches. Picking bags carelessly used are the cause of many bruised apples.

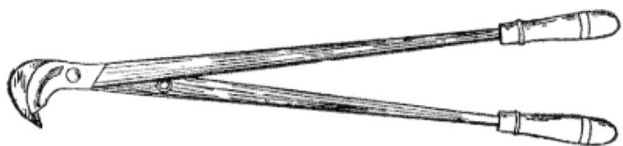

Figure 212.—Tree Pruners. The best made pruners are the cheapest. This long handled pruner is made of fine tool steel from the cutting parts clear to the outer ends of the wooden handles. A positive stop prevents the handles from coming together. Small one-hand pruning nippers are made for clean cutting. The blades of both pruners should work towards the tree trunk so the hook will mash the bark on the discarded portion of the limb.

The illustration shows one of the most convenient picking ladders. It is a double ladder with shelves to hold picking trays supported by two wheels and two legs. The wheels which are used to support one side of the frame are usually old buggy wheels. A hind axle together with the wheels works about right. The ladder frame is about eight feet high with ladder steps going up from each side. These steps also form the support for the shelves. Picking trays or boxes are placed on the shelves, so the latter will hold eight or ten bushels of apples, and may be wheeled directly to the packing shed if the distance is not too great.

Figure 213.—Shears. The first pair is used for sheep shearing. The second is intended for cutting grass around the edges of walks and flower beds.

164

Step-ladders from six to ten feet long are more convenient to get up into the middle of the tree than almost any other kind of ladder. Commercial apple trees have open tops to admit sunshine. For this reason, straight ladders are not much used. It is necessary to have ladders built so they will support themselves. Sometimes only one leg is used in front of a step-ladder and sometimes ladders are wide at the bottom and taper to a point at the top. The kind of ladder to use depends upon the size of the trees and the manner in which they have been pruned. Usually it is better to have several kinds of ladders of different sizes and lengths. Pickers then have no occasion to wait for each other.

FEEDING RACKS

Special racks for the feeding of alfalfa hay to hogs are built with slatted sides hinged at the top so they will swing in when the hogs crowd their noses through to get the hay. This movement drops the hay down within reach. Alfalfa hay is especially valuable as a winter feed for breeding stock. Sows may be wintered on alfalfa with one ear of corn a day and come out in the spring in fit condition to suckle a fine litter of pigs. Alfalfa is a strong protein feed. It furnishes the muscle-forming substances necessary for the young litter by causing a copious flow of milk. One ear of corn a day is sufficient to keep the sow in good condition without laying on too much fat. When shoats are fed in the winter for fattening, alfalfa hay helps them to grow. In connection with grain it increases the weight rapidly without adding a great deal of expense to the ration. Alfalfa in every instance is intended as a roughage, as an appetizer and as a protein feed. Fat must be added by the use of corn, kaffir corn, Canada peas, barley or other grains. Alfalfa hay is intended to take the place of summer pasture in winter more than as a fattening ration.

Figure 214.—Horse Feeding Rack. This is a barnyard hay feeder for horses and colts. The diagonal boarding braces each corner post and leaves large openings at the sides. Horses shy at small hay holes. The top boards and the top rail are 2 x 4s for strength. The bottom is floored to save the chaff.

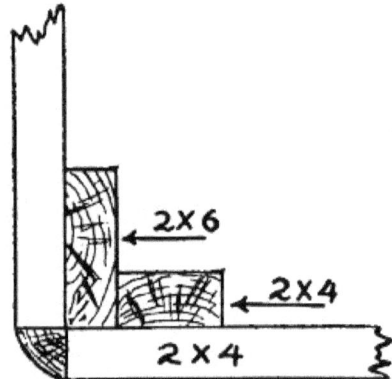

Figure 215.—Corner Post Detail of Horse Feeding Rack. A 2 x 6 is spiked into the edge of a 2 x 4, making a corner post 6″ across. The side boarding is cut even with the corner of the post and the open corner is filled with a two-inch quarter-round as shown.

166

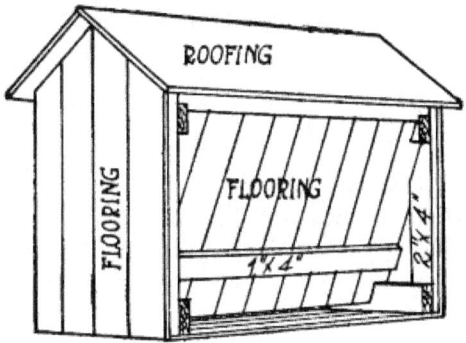

Figure 216.—Automatic Hog Feeder. The little building is 8′ x 12′ on the ground and it is 10′ high to the plates. The crushed grain is shoveled in from behind and it feeds down hopper fashion as fast as the hogs eat it. The floor is made of matched lumber. It should stand on a dry concrete floor.

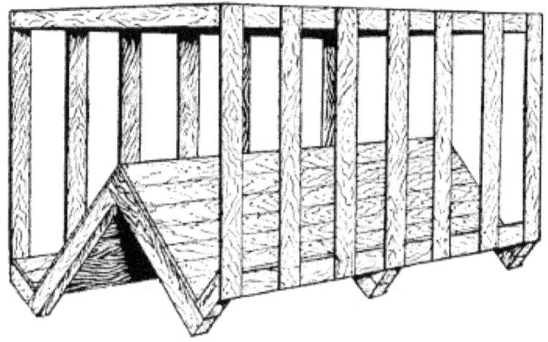

Figure 217.—Sheep Feeding Rack. The hay bottom and grain trough sides slope together at 45° angles. The boarding is made tight to hold chaff and grain from wasting.

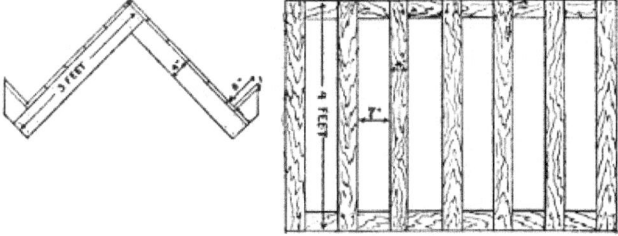

Figure 218.—Rack Base and Sides. The 2 x 4s are halved at the ends and put together at right angles. These frames are placed 3′ apart and covered with matched flooring. Light braces should be nailed across these frames a few inches up from the ground. The 1 x 4 pickets are placed 7″ apart in the clear, so the sheep can get their heads through to feed. These picketed frames are bolted to the base and framed around the top. If the rack is more than 9′ long there should be a center tie or partition. Twelve feet is a good length to make the racks.

SPLIT-LOG ROAD DRAG

The only low cost road grader of value is the split-log road drag. It should be exactly what the name implies. It should be made from a light log about eight inches in diameter split through the middle with a saw. Plenty of road drags are made of timbers instead of split logs, but the real principle is lost because such drags are too heavy and clumsy. They cannot be quickly adjusted to the varying road conditions met with while in use.

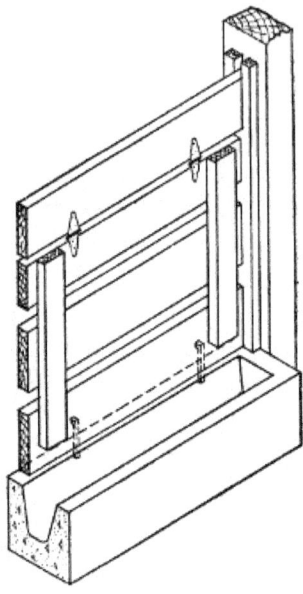

Figure 219.—Hog Trough. In a winter hog house the feed trough is placed next to the alley or passageway. A cement trough is best. A drop gate is hinged over the trough so it can be swung in while putting feed in the trough. The same gate is opened up level to admit hogs to the pen.

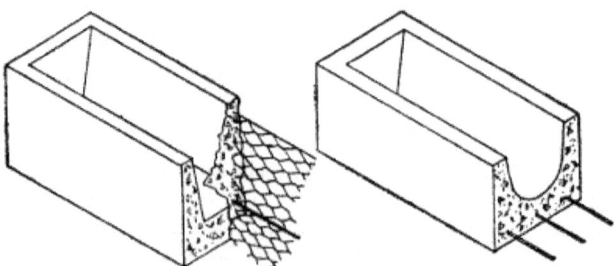

Figure 220.—Reinforced Hog Trough. The section of hog trough to the left is reinforced with chicken wire, one-inch mesh. The trough to the right is reinforced with seven $\frac{1}{4}''$ rods—three in the bottom and two in each side.

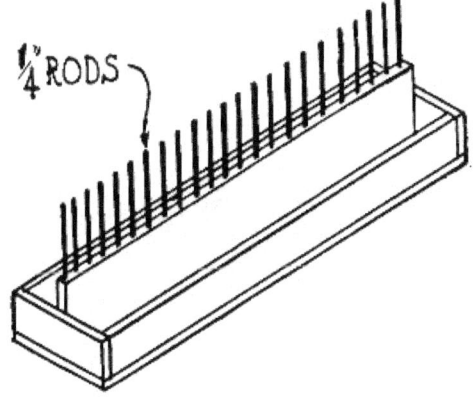

Figure 221.—Double Poultry Feeding Trough with Partition in the Center.

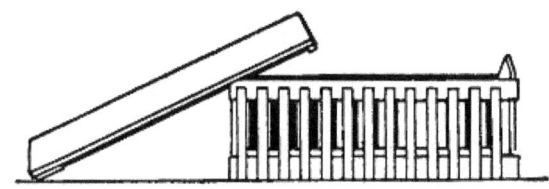

Figure 222.—Poultry Feeder with Metal or Crockery Receptacle.

The illustration shows the right way of making a road drag, and the manner in which it is drawn along at an angle to the roadway so as to move the earth from the sides towards the center, but illustrations are useless for showing how to operate them to do good work. The eccentricities of a split-log road drag may be learned in one lesson by riding it over a mile or two of country road shortly after the frost has left the ground in the spring of the year. It will be noticed that the front half of the road drag presents the flat side of the split log to the work of shaving off the lumps while the other half log levels and smooths and puddles the loosened moist earth by means of the rounded side. Puddling makes earth waterproof. The front, or cutting edge, is faced with steel. The ridges and humps are cut and shoved straight ahead or to one side to fill holes and ruts. This is done by the driver, who shifts his weight from one end to the other, and from front to back of his standing platform to distribute the earth to the best advantage. The rounded side of the rear half log presses the soft earth into place and leaves the surface smooth.

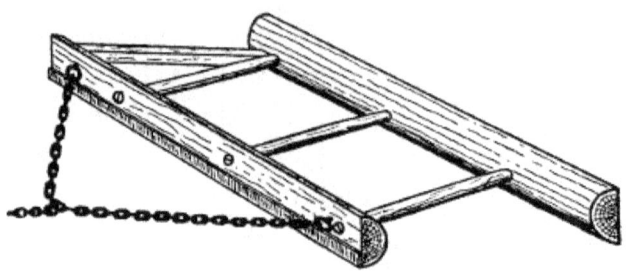

Figure 223.—Split-Log Road Drag. The front edge is shod with a steel plate to do the cutting and the round side of the rear log grinds the loosened earth fine and presses it into the wagon tracks and water holes.

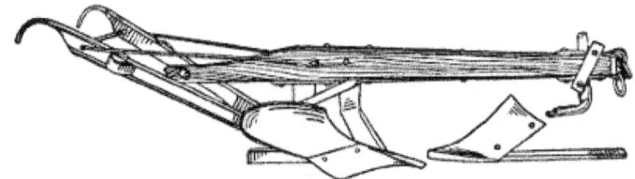

Figure 224.—Heavy Breaking Plow, used for road work and other tough jobs.

Unfortunately, the habit of using narrow tired wagons on country roads has become almost universal in the United States. To add to their destructive propensities, all wagons in some parts of the country have the same width of tread so that each wheel follows in paths made by other wheels, until they cut ruts of considerable depth. These little narrow ditches hold water so that it cannot run off into the drains at the sides of the roadway. When a rut gets started, each passing wheel squeezes out the muddy water, or if the wheel be revolving at a speed faster than a walk it throws the water, and the water carries part of the roadway with it so that small ruts are made large and deep ruts are made deeper. In some limited sections road rules demand that wagons shall have wide tires and have shorter front axles, so that with the wide tires and the uneven treads the wheels act as rollers instead of rut makers. It is difficult to introduce such requirements into every farm section. In the meantime the evils of narrow tires may be overcome to a certain extent by the persistent and proper use of the split-log road drag. These drags are most effectual in the springtime when the frost is coming out of the ground. During the muddy season the roads get worked up into ruts and mire holes, which, if taken in time, may be filled by running lengthwise of the road with the drag when the earth is still soft. When the ground shows dry on top and is still soft and wet underneath is the time the drags do the best work by scraping the drier hummocks into the low places where the earth settles hard as it dries.

A well rounded, smooth road does not get muddy in the summer time. Summer rains usually come with a dash. Considerable water falls in a short time, and the very act of falling with force first lays the dust, then packs the surface. The smooth packed surface acts like a roof, and almost before the rain stops falling all surface water is drained off to the sides so that an inch down under the surface the roadbed is as hard as it was before the rain. That is the reason why split log road drags used persistently in the spring and occasionally later in the season will preserve good roads all summer. It is very much better to follow each summer rain with the road drag, but it is not so necessary as immediate attention at the proper time in spring. Besides, farmers are so busy during the summer months that they find it difficult to spend the time. In some sections of the middle West one man is hired to do the dragging at so much per trip over the road. He makes his calculations accordingly and is prepared to do the dragging at all seasons when needed. This plan usually works out the best because one man then makes it his business and he gets paid for the amount of work performed. This man should live at the far end of the road division so that he can smooth his own pathway leading to town.

STEEL ROAD DRAG

Manufacturers are making road drags of steel with tempered blades adjustable to any angle by simply moving the lever until the dog engages in the proper notch. Some of these machines are made with blades reversible, so that the other side can be used for cutting when the first edge is worn. For summer use the steel drag works very well, but it lacks the smoothing action of a well balanced log drag.

SEED HOUSE AND BARN TRUCKS

Bag trucks for handling bags of grain and seeds should be heavy. Bag truck wheels should be eight inches in diameter with a three-inch face. The steel bar or shoe that lifts and carries the bag should be twenty-two inches in length. That means that the bottom of the truck in front is twenty-two inches wide. The wheels run behind this bar so the hubs do not project to catch against standing bags or door frames. The length of truck handles from the steel lift bar to the top end of the hand crook is four feet, six inches. In buying bag trucks it is better to get the heavy solid kind that will not upset. The light ones are a great nuisance when running them over uneven floors. The wheels are too narrow and too close together and the trucks tip over under slight provocation. Platform trucks for use in moving boxes of apples or crates of potatoes or bags of seed in the seed house or warehouse also should be heavy. The most approved platform truck, the kind that market men use, is made with a frame four feet in length by two feet in width. The frame is made of good solid hardwood put together with mortise and tenon. The cross pieces or stiles are three-quarters of an inch lower than the side pieces or rails, which space is filled with hardwood flooring boards firmly bolted to the cross pieces so they come up flush with the side timbers. The top of the platform should be sixteen inches up from the floor. There are two standards in front which carry a wooden crossbar over the front end of the truck. This crossbar is used for a handle to push or pull the truck. The height of the handle-bar from the floor is three feet. Rear wheels are five inches in diameter and work on a swivel so they turn in any direction like a castor. The two front wheels carry the main weight. They are twelve inches in diameter with a three-inch face. The wheels are bored to fit a one-inch steel axle and have wide boxings bolted to the main timbers of the truck frame. Like the two-wheel bag truck, the wheels of the platform truck are under the frame so they do not project out in the way, which is a great advantage when the truck is being used in a crowded place.

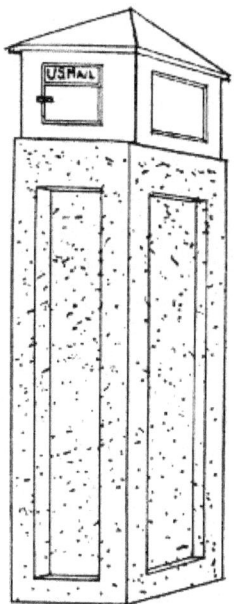

Figure 226.—Farm Gate Post with Copper Mail Box.

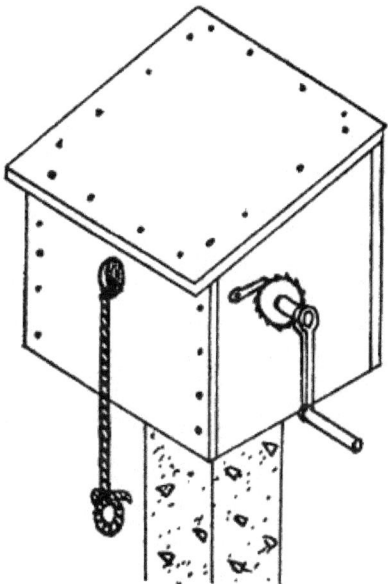

Figure 227.—Concrete Post Supporting a Waterproof Clothes Line Reel Box.

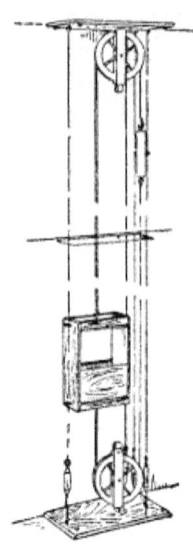

Figure 228.—Dumb Waiter. The cage is poised by a counterweight. It is guided by a rope belt which runs on grooved pulleys at the top and bottom.

HOME CANNING OUTFIT

There are small canning outfits manufactured and sold for farm use that work on the factory principle. For canning vegetables, the heating is done under pressure because a great deal of heat is necessary to destroy the bacteria that spoil vegetables in the cans. Steam under pressure is a good deal hotter than boiling water. There is considerable work in using a canning outfit, but it gets the canning out of the way quickly. Extra help may be employed for a few days to do the canning on the same principle that farmers employ extra help at threshing time and do it all up at once. Of course, fruits and vegetables keep coming along at different times in the summer, but the fall fruit canning may be done at two or three sittings arranged a week or two apart and enough fruit packed away in the cellar to last a big family a whole year. Canning machinery is simple and inexpensive. These outfits may be bought from $10 up. Probably a $20 or $25 canner would be large enough for a large family, or a dozen different families if it could be run on a co-operative plan.

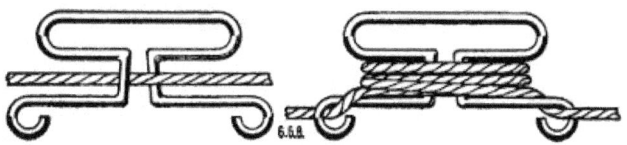

Figure 229.—Clothes Line Tightener. This device is made of No. 9 wire bent as shown in the illustration.

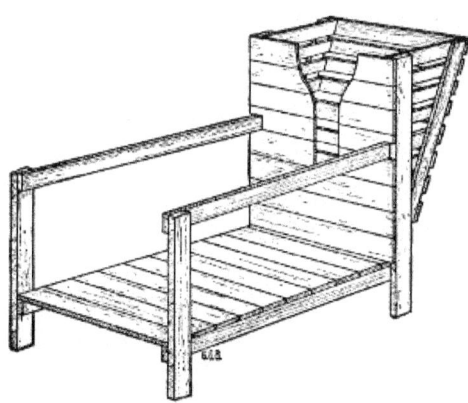

Figure 230.—Goat Stall. Milch goats are milked on a raised platform. Feed is placed in the manger. The opening in the side of the manger is a stanchion to hold them steady.

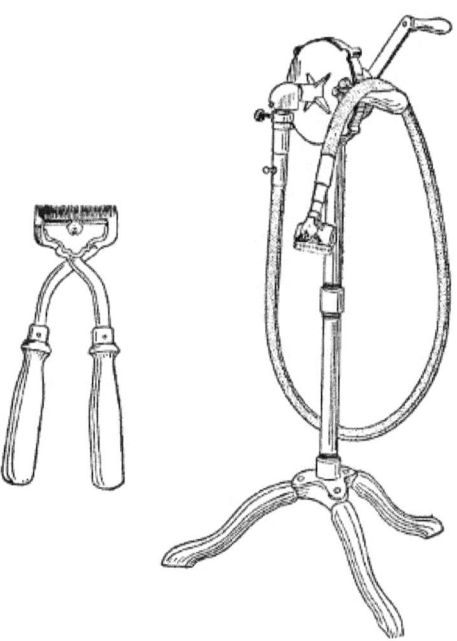

Figure 231.—Horse Clippers. Hand clippers are shown to the left. The flexible shaft clipper to the right may be turned by hand for clipping a few horses or shearing a few sheep, but for real business it should be driven by an electric motor.

ELECTRIC TOWEL

The "air towel" is sanitary, as well as an economical method of drying the hands. A foot pedal closes a quick-acting switch, thereby putting into operation a blower that forces air through an electric heating device so arranged as to distribute the warmed air to all parts of the hands at the same time. The supply of hot air continues as long as the foot pedal is depressed. The hands are thoroughly dried in thirty seconds.

STALLS FOR MILCH GOATS

Milch goats are not fastened with stanchions like cows. The front of the manger is boarded tight with the exception of a round hole about two feet high and a slit in the boards reaching from the round opening to within a few inches of the floor. The round hole is made large enough so that the goat puts her head through to reach the feed, and the slit is narrow enough so she cannot back up to pull the feed out into the stall. This is a device to save fodder.

Figure 232.—Hog Catching Hook. The wooden handle fits loosely into the iron socket. As soon as the hog's hind leg is engaged the wooden handle is removed and the rope held taut.

STABLE HELPS

Figure 233.—Bull Nose-Chain. Cross bulls may be turned out to pasture with some degree of safety by snapping a chain like this into the nose-ring. The chain should be just long enough to swing and wrap around the bull's front legs when he is running. Also the length is intended to drag the ring where he will step on it with his front feet. There is some danger of pulling the nose ring out.

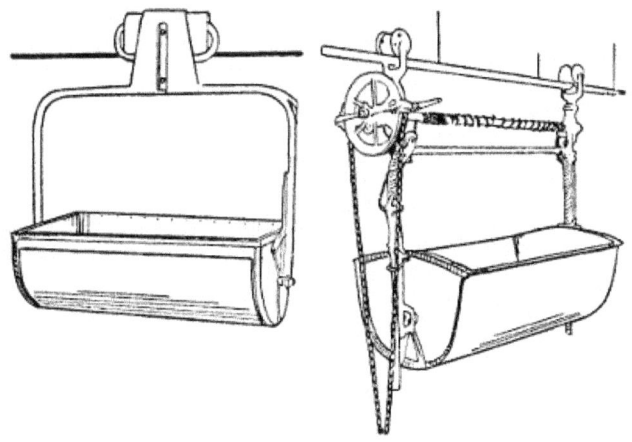

Figure 234.—Manure Carriers. There are two kinds of manure carriers in general use. The principal difference is the elevator attachment for hoisting when the spreader stands too high for the usual level dump.

Overhead tracks have made feed carriers possible. Litter or feed carriers and manure carriers run on the same kind of a track, the only difference is in size and shape of the car and the manner in which the contents are unloaded. Manure carriers and litter carriers have a continuous track that runs along over the manure gutters and overhead lengthwise of the feed alleys. There are a number of different kinds of carriers manufactured, all of which seem to do good service. The object is to save labor in doing the necessary work about dairy stables. To get the greatest possible profit from cows, it is absolutely necessary that the stable should be kept clean and sanitary, also that the cows shall be properly fed several times a day. Different kinds of feed are given at the different feeding periods. It is impossible to have all the different kinds of food stored in sufficient quantities within easy reach of the cows. Hence, the necessity of installing some mechanical arrangement to fetch and carry. The only floor carrier in use in dairy stables is a truck for silage. Not in every stable is this the case. Sometimes a feed carrier is run directly to the silo. It depends a good deal on the floor what kind of a carrier is best for silage. The advantage of an overhead track is, that it is always free from litter. Where floor trucks are used, it is necessary to keep the floor bare of obstruction. This is not considered a disadvantage because the floor should be kept clean anyway.

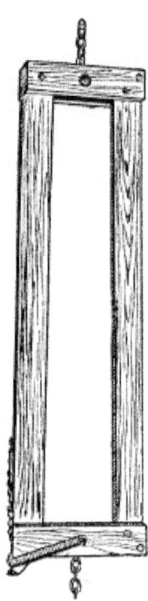

Figure 235.—Cow Stanchion. Wooden cow stanchions may be made as comfortable for the cows as the iron ones.

HOUSE PLUMBING

When water is pumped by an engine and stored for use in a tank to be delivered under pressure in the house, then the additional cost of hot and cold water and the necessary sink and bath room fixtures is comparatively small. Modern plumbing fixtures fit so perfectly and go together so easily that the cost of installing house plumbing in the country has been materially reduced, while the dangers from noxious gases have been entirely eliminated. Open ventilator pipes carry the poisonous gases up through the roof of the house to float harmlessly away in the atmosphere. Septic tanks take care of the sewerage better than the sewer systems in some towns. Plumbing fixtures may be cheap or expensive, according to the wishes and pocketbook of the owner. The cheaper grades are just as useful, but there are expensive outfits that are very much more ornamental.

FARM SEPTIC TANK

Figure 236.—Frame for Holding Record Sheets in a Dairy Stable.

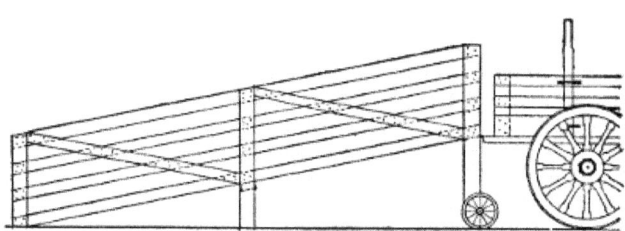

Figure 237.—Loading Shute for Hogs. This loading shute is made portable and may be moved like a wheelbarrow.

Supplying water under pressure in the farmhouse demands a septic tank to get rid of the waste. A septic tank is a scientific receptacle to take the poison out of sewerage. It is a simple affair consisting of two underground compartments, made water-tight, with a sewer pipe to lead the waste water from the house into the first compartment and a drain to carry the denatured sewerage away from the second compartment. The first compartment is open to the atmosphere, through a ventilator, but the second compartment is made as nearly air-tight as possible. The scientific working of a septic tank depends upon the destructive work of two kinds of microscopic life known as aerobic and anaerobic forms of bacteria. Sewerage in the first tank is worked over by aerobic bacteria, the kind that require a small amount of oxygen in order to live and carry on their work. The second compartment is inhabited by anaerobic bacteria, or forms of microscopic life that work practically without air. The principles of construction require that a septic tank shall be large enough to contain two days' supply of sewerage in each compartment; thus, requiring four days for the sewerage to enter and leave the tank.

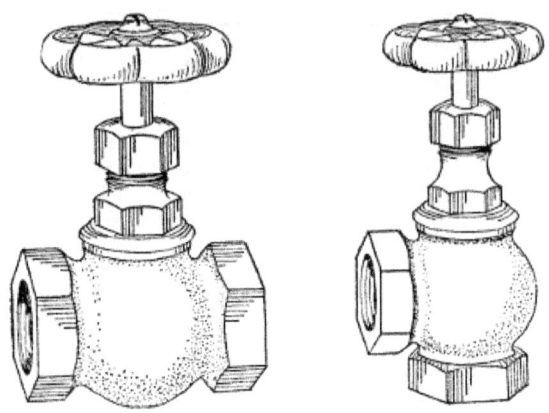

Figure 238.—Brass Valves. Two kinds of globe valves are used in farm waterworks. The straight valve shown to the left and the right angle valve to the right. Either one may be fitted with a long shank to reach above ground when pipes are laid deep to prevent freezing.

Estimating 75 gallons daily of sewerage for each inhabitant of the house and four persons to a family, the septic tank should be large enough to hold 600 gallons, three hundred gallons in each compartment, which would require a tank about four feet in width and six feet in length and four feet in depth. These figures embrace more cubic feet of tank than necessary to meet the foregoing requirements. It is a good plan to leave a margin of safety.

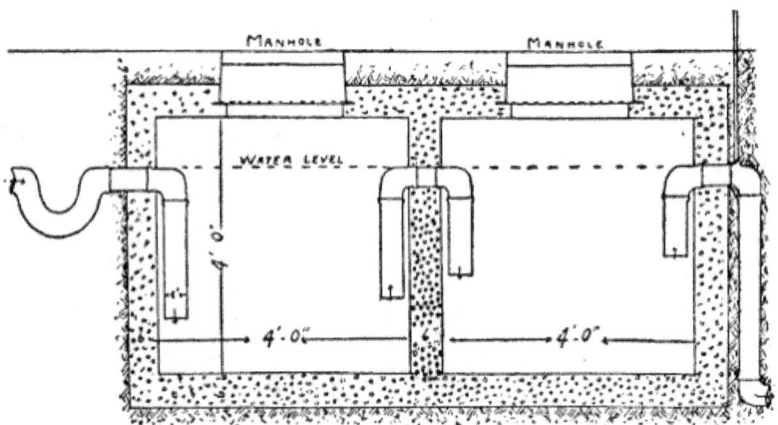

Figure 239.—Septic Tank, a double antiseptic process for purifying sewerage.

It is usual to lay a vitrified sewer, four inches in diameter, from below the bottom of the cellar to the septic tank, giving it a fall of one-eighth inch in ten feet. The sewer enters the tank at the top of the standing liquid and

180

delivers the fresh sewerage from the house through an elbow and a leg of pipe that reaches to within about six inches of the bottom of the tank. The reason for this is to admit fresh sewerage without disturbing the scum on the surface of the liquid in the tank. The scum is a protection for the bacteria. It helps them to carry on their work of destruction. The same principle applies to the second compartment. The liquid from the first compartment is carried over into the second compartment by means of a bent pipe in the form of a siphon which fills up gradually and empties automatically when the liquid in the first compartment rises to a certain level. The discharging siphon leg should be the shortest. The liquid from the second compartment is discharged into the drain in the same manner. There are special valves made for the final discharge, but they are not necessary. The bottom of the tank is dug deep enough to hold sewerage from two to four feet in depth. The top surface of the liquid in the tank is held down to a level of at least six inches below the bottom of the cellar. So there is no possible chance of the house sewer filling and backing up towards the house. Usually the vitrified sewer pipe is four inches in diameter, the septic tank siphons for a small tank are three inches in diameter and the final discharge pipe is three inches in diameter, with a rapid fall for the first ten feet after leaving the tank.

Septic tanks should be made of concrete, waterproofed on the inside to prevent the possibility of seepage. Septic tank tops are made of reinforced concrete with manhole openings. Also the manhole covers are made of reinforced concrete, either beveled to fit the openings or made considerably larger than the opening, so that they sit down flat on the top surface of the tank. These covers are always deep enough down in the ground so that when covered over the earth holds them in place.

In laying vitrified sewer it is absolutely necessary to calk each joint with okum or lead, or okum reinforced with cement. It is almost impossible to make a joint tight with cement alone, although it can be done by an expert. Each length of the sewer-pipe should be given a uniform grade. The vitrified sewer is trapped outside of the building with an ordinary S-trap ventilated, which leaves the sewer open to the atmosphere and prevents the possibility of back-pressure that might drive the poisonous gases from the decomposing sewerage through the sewer back into the house. In this way, the septic tank is made entirely separate from the house plumbing, except that the two systems are connected at this outside trap.

It is sometimes recommended that the waste water from the second compartment shall be distributed through a series of drains made with three-inch or four-inch drain tile and that the outlet of this set of drains shall empty into or connect with a regularly organized field drainage system. Generally speaking, the final discharge of liquid from a septic tank that is properly constructed is inoffensive and harmless. However, it is better to use every possible precaution to preserve the health of the family, and it is better to dispose of the final waste in such a way as to prevent any farm animal from drinking it.

While manholes are built into septic tanks for the purpose of examination, in practice they are seldom required. If the tanks are properly built and rightly proportioned to the sewerage requirements they will take care of the waste water from the house year after year without attention. Should any accidents occur, they are more likely to be caused by a leakage in the vitrified sewer than from any other cause. Manufacturers of plumbing supplies furnish the siphons together with instructions for placing them properly in the concrete walls. Some firms supply advertising matter from which to work out the actual size and proportions of the different compartments and all connections. The making of a septic tank is simple when the principle is once understood.

www.ingramcontent.com/pod-product-compliance
Lightning Source LLC
Chambersburg PA
CBHW071428180526
45170CB00001B/256